# ACTIVE MEASURES

Активные
мероприятия

# ACTIVE MEASURES

Spring 2013 · Volume II

# ACTIVE MEASURES

Spring 2013 · Volume II





*Inquiries should be directed to journal.activemeasures@gmail.com, or:*

Active Measures, *c/o The Institute of World Politics*

*1521 16th Street NW*

*Washington, D.C. 20036*

# From the Provost

J. Michael Waller

What do the following issues have in common?

- ☑ the Iranian navy
- ☑ the Chinese way of war
- ☑ integration of cultural geography with psychological operations
- ☑ Soviet subversion in higher education and Hollywood
- ☑ a dispute about Israeli farms in territory claimed by Lebanon
- ☑ U.S. policy toward outer space

Their main common trait is that they are discussed in this issue of *Active Measures*.

But readers of *Active Measures* don't just look at surface traits. They look deeper. They think strategically, in an integrated fashion. They tie together seemingly

*__J. Michael Waller__ is the Vice President and Provost of The Institute of World Politics. He holds the Walter and Leonore Annenberg Chair in International Communication, and directs the Institute's graduate programs on public diplomacy and political warfare. He has been a scholar-practitioner in public diplomacy, political warfare, psychological operations and information operations in support of U.S. foreign and military policy for more than 25 years. He was a member of the staff of the U.S. House of Representatives and the U.S. Senate, served on the White House Task Force on Central America, was an operative for members of the White House Active Measures Working Group, and has been a consultant to the Senate Foreign Relations Committee, the U.S. Information Agency, the U.S. Agency for International Development, the Office of the Secretary of Defense, the U.S. Army, and the U.S. Special Operations Command. In 2006 he received a citation from the Director of the FBI for "exceptional service in the public interest." He has designed and conducted specialized training courses for the U.S. military and civilian agencies.*

unrelated pieces to get a global perspective.

The authors of each of the articles didn't intend for them to link to one another. They didn't even know about the other pieces.

So how could they possibly fit together?

Read them like a strategist and explore beyond the obvious. Perhaps the Iranian navy and Iranian-backed forces in Lebanon, both with Chinese-made weapons, could be carrying out elements of the Chinese way of war, while US policy toward outer space, which guarantees American military superiority and global information dominance, holds the key to successful integration of cultural geography with psychological operations to provide non-kinetic, human-centric solutions to the regimes that plague us in Teheran and Beijing. The only constraints to the US arriving at such peaceful solutions to the apparently inevitable conflicts with Iran and China are the legacies of weakened national identity and self confidence held over from decades of Soviet subversion of American higher education and popular culture.

Too farfetched? How about this combination: With English-speaking Chinese outnumbering Americans, China applies its way of war toward integrating cultural geography and psychological operations for world dominance without open warfare. Beijing learns lessons from Soviet subversion of American higher education and entertainment to weaken the US psychologically, over time persuading the United States to give up its dominance of outer space, while arming proxy forces in Iran and Lebanon to attack the US indirectly by hitting its interests in the Middle East.

The combinations for discussion are limited only by one's imagination.

Like any pieces of intelligence, the articles might conceivably link together – or they might have nothing to do with one another whatsoever.

The point is that the strategic thinker will seek to integrate all available information to look at variables that hide beneath the surface, to stay ahead of all opposition, to have the prudence to divine the realistic from the unrealistic, and to be courageous enough to stand up for one's convictions.

We're proud to bring you another issue of our students' own journal, *Active Measures*. Read it and tell us what you think: Do these articles have any connection to one another, and if so, how?

# From the Editors

In the foreward of this edition, Dr. Waller challenged you, the reader, to look at this article from a strategic viewpoint. He challenged you to discover the possible connections between the articles, and in that way, stay ahead of the opposition. This is a difficult task given that the articles vary widely and were written without the intention of being linked together. But the fact remains that, in today's globalized world, when one nation, NGO, or any other international organization does something, the vibrations are felt far outside of the originator's borders. The information revolution has assured that groups that are separated by miles can act in concert with one another in order to achieve their strategic objectives for good or ill. Often, the connections between states, NGO's, MNCs, etc. are not easily discernible, but they do exist and they will have an impact on the security environment. The practitioner of statecraft must see these connections before they happen so that in dealing with the consequences of an action all possibilities of reverberation are accounted for and addressed.

The United States currently has little formal mechanisms for creating and following through with an integrated strategy that takes into account the skills and interests of all departments. The National Security Council exists partially to ensure that the cabinet secretaries, and others as required, agree and support the President's policies. But more is needed. In today's budget environment, the US must not duplicate programs or allow Departments to act in ways that undermine each other. The USG must act in concert, utilizing all tools of statecraft, to achieve our objectives. This means on a very basic level that the right hand and the left hand must communicate and even assist each other.

The United States must integrate into its strategy techniques that have long been forgotten since the Cold War - public diplomacy, psychological operations, information operations, political warfare, etc. These tools were necessary to win the Cold War and they are necessary to win the ideological battles we face today. An increased number of drones or carriers will never convince Islamists that

America is not the Great Satan or the PRC that communism still cannot defeat us. The challenges the US faces today are complex and interconnected in ways that are often hidden from plain site. This simply means that we must look deeper and see these connections before they surprise us. Our responses to today's challenges must be integrated and we must utilize all tools available to us. This is the only way we can truly achieve our strategic goals in today's globalized, interconnected word.

# From the North Caucasus to Boston: Globalization of the Caucasus Emirate

Olivia Goldstein

Monday April 15th, 2013: 27,000 runners participated in the annual Boston Marathon. As spectators gathered to watch the runners take part in the world's oldest marathon, explosions went off near the finish line turning this celebration into a scene of chaos and destruction. The homemade explosions left three innocent people dead and nearly 150 severely injured. The culprits behind the somber and tragic event were two brothers with ethnic ties to Chechnya.

The Tsarnaev brothers identified and embraced their Chechen nationality. However over the course of the past few years, multiple indicators suggested the brothers embraced radical Islam. The radical Islam coupled together with Chechen pride indicates a possible linkage to the al-Qaeda supported terror group the Caucasus Emirate. However, first the history of Chechen insurgence needs to be examined.

The Chechens have generations of rebellious spirit. Their home has high mountain ranges with thick beech forests: a perfect environment for guerilla warfare. Chechnya has numerous shrines and monuments dedicated to their resistance against Russian rule. However, Islamic extremists quickly exploited the Chechen

***Olivia Goldstein****, from Buffalo, NY, is a Master of Art degree recipient of The Institute of World Politics. She earned a Bachelor of Art in Economics and History and a Certificate of Russian and East European Studies from the University of Pittsburgh. Olivia's academic and professional work focuses on Russian foreign and domestic policy.* *She would like to thank her supportive family and friends; the faculty at IWP; and U.S. Department of State Foreign Service Officer Michelle Yerkin, who serves as Olivia's mentor.*

insurgents. Primarily, the main threat in the region is the terrorist group known as the Caucasus Emirate.

Islam remains the largest cultural difference between the North Caucasus region and Russia. The Chechens are largely Muslim not living under sharia law.[1] When the Soviet Union fell it was discovered that many Chechens did not know how to pray and regularly consumed alcohol. Clearly Chechens did not strictly follow the rules of Islam. Thus, initially Islam was not the primary motivation for separation between Chechnya and Russia.

Historically, relations between Chechnya and Russia have been problematic since the eighteenth century. As stated previously, Russia's Muslim community in the North Caucasus was isolated. The historical tension and isolated Muslim community provided a fertile ground for ultimate radicalization. Therefore radical Islam succeeded in Chechnya and the North Caucasus because the inhabitants did not know what Islam was. The more they learned about the religion the more committed they became until ultimately recruited.[2] Unfamiliarity with Islam coupled with Russian oppression led to radicalized Islam in Chechnya.

However, what began as a nationalist separation movement soon morphed into an international Islamist terror group and global terrorist attacks. The North Caucasus was a particularly strategic chokepoint for Jihadists because of its close proximity to both Russia and Europe. The goal was to transform Chechnya from its dated pre-Islamic structure to an Islamic-Jihadist structure and abandon previous known culture.[3] As in the case with many insurgencies, the organizers' goals developed over time. There is great controversy surrounding how the Chechens turned from nationalist separatists to a global terror network associated with al-Qaeda. Ultimately with the influence of Middle-Eastern Islamist-Jihadists, Chechen rebels developed jihadist ideology. In 2001, Ayman al-Zawahiri proclaimed in the Knights under the Banner of the Prophet, "to gather around the fighting states and support them...our Islamist movement and its avant-garde of the jihad fighters, together with the whole umma, must force the major criminals...into a fight." He then specifically references Afghanistan and Chechnya as "the true capital of Islam," and thus it is "the highest of all duties to defend both countries through word, action, and consultation."[4] Quite explicitly here, the current leader of al-Qaeda called on jihadists to assist the Chechens.

When the Soviet Union fell, the Chechens attempted separation. Moscow dismissed their initial declaration of independence as inane. By 1994, Moscow realized the severity of losing Chechnya and subsequently Russian troops entered Chechnya. As the Chechen war progressed, so did the Chechen support system. In the Islamist world, the Chechen insurgency gained legitimacy as a jihad. Jihadists sympathized with the struggling Chechens against their oppressor, Russia, and agreed that the North Caucasus was Islamic territory.[5]

Furthermore, since the Islamic political revival began in the 1970s, there became an increased salience of religion identity. In this particular case, the Muslim community was extremely sympathetic toward the Chechens because of the Afghan struggle against the Soviets in the 1980s.[6] While the Taliban was in power, they offered training camps to Chechen rebels and encouraged Islamic militants in Central Asia. Many militants even volunteered to fight alongside the Taliban against NATO.[7]

Terrorism during the 2000s grew to an unprecedented level in Russia. In October 2002 Chechen terrorists held 850 Russians hostage in Moscow's Dubrovka Theater. Among the suicide squad were numerous females wearing vests with 3-5 kilograms of homemade explosives.[8] After three days, Russian special-forces pumped chemicals into the theater and regained control of the theater. However, Russian authorities failed to inform doctors what the chemical was that knocked out the hostages and terrorists. Thus, 129 hostages died due to the deadly gas.[9] Worth noting here is the immoral counter-terrorism techniques of the Russian government, but it is not the focus of this research.

After the incident at the Dubrovka Theater, women suicide bombers became a regular strategy for the Chechens. Widowed and alone many women volunteered to sacrifice themselves for Allah and an independent Chechnya. By participating in suicide bombings, many women hoped they could inflict pain on the Russians as they endured. Shahidka, or Black Widows as they became to be known, were a new strategy implemented by the Chechens. In the case of 23-year old Zarema Muzhikhoyeva, she surrendered after attempting to blow herself up in central Moscow on July 9th, 2003. Upon being interrogated, she admitted that her grandparents had brought her up after being abandoned by her parents. She married young and got pregnant early. Her husband died and her daughter was taken away by her in-laws. As a widow and single mother, she was unlikely of ever finding work to support herself. She contacted a woman close to the jihadists and volunteered herself as a suicide bomber in exchange for $1000 to help raise her daughter.[10] Young Chechen women randomly bombed busy areas of Moscow, including the 2010 Moscow metro bombing. At present, the Tsarnaevs' mother is suspected of having a role in the Boston bombings. Potentially, Zubeidat Tsarnaeva discussed jihad and radical ideas with her now deceased son over the telephone. While it is still early in the investigation, it is worth noting Zubeidat's influence and role in the 2013 Boston Marathon bombings.

Another major terrorist attack in Russia occurred in 2004 at a school in Beslan. A group of Jihadists seized a school for three days holding over 1,100 people hostage, a majority of whom were children.[11] After three days, Russian SWAT teams entered the school with heavy artillery. Russian forces attacked the school with flamethrowers and grenade launchers and killed 332 people and children.[12] The crisis finally ended when the terrorists triggered the bombs in the gymnasium bringing down the roof and killing remaning of the hostages. The ultimate goal of the Beslan strike however was not to further the Chechen rebellion. Instead, a captured terrorist admitted the goal was to further the Islamist jihad throughout the Caucasus. Unfortunately, the Russian government refused negotiations to end the Beslan crisis. Therefore, once again Russian oppression laid the foundation for radicalization of the Chechen insurgents.

Strikingly different from the beginning separatist movement, now Chechen insurgents are jihadists.[13] The international community criticized Russian counterterrorism and counterinsurgency operations. Throughout the bloody insurgency, the Russian government systemically violated human rights. Russian forces brutally raped women, tortured civilians and insurgents alike, executed thousands, all while making others vanish or disappear. Russian forces razed villages and Grozny killing thousands of people. Authorities paid little attention to the ideological basis of the North Caucasus predicament and responded with sheer brute violence. As a result, Russian policy radicalized the Chechens and provided greater "legitimacy" to the jihadists working to exploit the Chechen cause.

In retrospect, why did Moscow not strike a deal with Chechnya in the early 1990s? Had they come to an agreement on economic and cultural autonomy, thousands of lives and damage could have been prevented. Yet, the Kremlin has a history of imperialist arrogance. For Moscow, it was not an option to allow Chechnya to secede because of its strategic location. Mainly, a strategic oil pipeline runs from Azerbaijan through Dagestan and Chechnya into the Russian port Novorossiisk. The constant power struggle for the domination over and gas transportation routes from the Caspian Sea made the conflict nearly inevitable unless one side was willing to compromise.

October 31st, 2007 the Caucasus Emirate officially came into existence as the successor to the Chechen Republic of Ichkeria. Emir Doku Umarov leads the self-proclaimed virtual state entity. In the official declaration of the Caucasus Emirate, Umarov declared, "I reject all kafir laws established in the world. I reject all laws and systems established by the infidels in the land of the Caucasus. (The task is) to expel the infidels from the Caucasus (and to make the territory) the house of peace. (In addition,) we must re-conquer all the historical land of Muslims, beyond the boundaries of Caucasus." Umarov classified all non-Muslims as infidels and "objects of destruction." In his *fatwahs*, Umarov asserted that anyone fighting Muslims anywhere in the world is the Caucasus Emirate's enemy. He persuaded followers to shoot down the infidels.[14]

The Caucasus Emirate is committed to the global confrontation between the

Muslim community and its enemies worldwide. This much is obvious from their website Kavkaz Center. The site is available in Arabic, Russian, English, Ukranian, and Turkish. Clearly, their cyber presence and influence compensates for an arguable lack of physical presence. Similar to al-Qaeda, the Caucasus Emirate relies on homegrown terrorists to carry out isolated acts of terror.

Since its inception, the Caucasus Emirate claimed responsibility for at least three major attacks in Russia. In November 2009, a high speed train traveling from Moscow to St. Petersburg was bombed 250 miles outside of Moscow. At the time of the derailment, the Nevsky Express carried nearly 660 passengers. The bombing caused about 30 deaths and nearly 130 injuries. Shortly after, Russian forces raided the village Ekazhevo and killed eight suspected militants and detained ten others. The Russian courts found four of those detained guilty of murder, organizing an act of terrorism, and illegal arms trafficking thereby jailing them for life.[15]

With 185 stations and 12 lines, there is not doubt that the Moscow metro is the world's second most heavily used rapid transit system. On an average day, roughly 6.5 million people use the metro rail. Therefore, if an attack occurred during peak rush hours one can only imagine the damage and destruction it could cause. On March 29, 2010 suicide bombers attacked Lubyanka and Park Kultry metro stops. The first bomb detonated at 07:56 a.m. just as the train doors opened killing 15 people on the train and 11 on the platform of Lubyaka with 1.5 kilograms of trinitrotoluene (TNT). The second explosion went off at 08:38 a.m. killing 14 at Park Kultry station with two kilograms of TNT. Following the attack, Umarov publicly acknowledged responsibility and announced more attacks would follow.

Less than one year later on January 24th, 2011 a suicide bomber attacked the international arrival hall of Moscow's busiest airport, Domodedovo. The attack on Domodedovo left 36 people dead and an additional 180 injured. Once again, Umarov ordered the attack via Kavkaz Center. Shortly after the Domodedovo bombing, the United Nations finally recognized the sobering threat of the Umarov and the Caucasus Emirate. On March 10, 2011, the United Nations listed Doku Umarov as an international terrorist specifically associated with al-Qaeda.[16] Umarov publicly describes Western countries as the enemies of all Muslims. His goals are to install *sharia* law across the Caucasus Emirate, sounding strikingly similar to al-Qaeda rhetoric.

At this point the Caucasus Emirate has not claimed responsibility behind the Boston Marathon bombing. In addition, authorities have not linked Tamerlan and Dzhokhar Tsarnaev to the Caucasus Emirate. However, it is worth exploring how the Caucasus Emirate exploits and brainwashes younger generations of Chechen Muslims. While there has been some tension in upper leadership, including Umarov stepping down in August 2010 only to retract his statement the following day, younger generations of North Caucasian jihadists continue to support Umarov. In addition, many students from the Northern Caucasus are sponsored by wealthy Muslim elites to study in the Faculty of Shari'a and Law, the Faculty of Islamic 'Propaganda' and Theology at the University of al-Azhar.[17]

Coming full circle, what exactly is the relationship between the Tsarnaev brothers and terrorism, more specifically the Caucasus Emirate? The Boston Marathon bombings provoked American fears of terrorism among seemingly normal youths. The first fact worth exploring is the Tsarnaev brothers' strong Chechen identity. The elder brother's, Tamerlan, connection is better established. Tamerlan made statements about his support of Chechen independence in Russia. He was also a devout Muslim. As early as 2011, Tamerlan sent text messages to family members suggesting his willingness to die for Islam. Furthermore in 2011, the Russian government alerted U.S. intelligence of Tamerlan's association with extremist groups. Unfortunately, no probable cause was found and Tamerlan slipped away. In addition, just last year Tamerlan visited Dagestan for an extended period of time leaving his family in the United States. As most Americans agree, terrorism is bred in failed states. As Paul Goble pointed out in his testimony before the House Subcommittee on Europe, Eurasia, and Emerging Threats and the Subcommittee of Terrorism, Nonproliferation, and Trade, Acting President of Dagestan (Ramazan Abdulatipov) described his republic as a failed state.[18] Therefore, the odds of Tamerlan visiting Dagestan for six months and not having exposure to Islamic extremists is very slim.

After six months in Dagestan, Tamerlan returned and created a YouTube account. Tamerlan's YouTube page is the most concrete source connecting him to the Caucasus Emirate.[19] His page includes videos on "Islam" and "Terrorism." In addition, his user name "muaz[20] seyfullah" could be a reference to Emir Seyfullah, Supreme Qadi of the Caucasus Emirate.

Tamerlan's younger brother, Dzhokhar, has less established connections to Islamist extremism. By most accounts, he was a reasonably well-adjusted American immigrant. Merely 19 years old, Dzhokar attended the University of Massachusetts-Dartmouth. Classmates described Dzhokar as friendly, sociable and nothing less than a typical college student. Many of those familiar with the Tsarnaevs are convinced that Tamerlan brainwashed Dzhokar. As the investigation proceeds authorities hopefully will discover what truly motivated Dzhokar Tsarnaev.

As some may argue, the Chechen heritage is irrelevant at this point in the investigation. These reactionary accusations suggest that the Tsarnaev brothers were merely self-radicalized, deranged individuals and the isolated Boston Marathon incident lacked the tradecraft of a terror organization. This simple-minded view disregards the clear connections the brothers have to terrorist groups in the North

Caucasus. The homemade bombs closely resembled instructions from an article in the al-Qaeda magazine *Inspire*. The article "How to Build a Bomb in the Kitchen of Your Mom" in the first issue of *Inspire* provides instructions on how to construct bombs with kitchen pressure cookers. The magazine article provided the blueprints for the weapons used in the attack on Boston.

Many terrorist groups including al-Qaeda and the Caucasus Emirate rely on a new tactic of homegrown terrorism. This political warfare tool targets American and other Western citizens and encourages individual jihadist attacks. However, the Tsarnaev brothers were not disenchanted "lone-wolves" such as Adam Lanza or James Holmes.[21] As previously stated, the brothers were well-adjusted immigrants. Furthermore, Tamerlan's six-month mysterious trip to Dagestan clearly suggests he received both training and support for carrying out the attack in Boston. Both brothers followed their roots in Chechnya and Islam and carried out the premeditated attack under the guidance and mentoring of *Misha*.[22] The Tsarnaevs clearly had ideological links to the Caucasus Emirate. The point of dispute is whether they attended some type of militant training abroad leading to operational links to the Caucasus Emirate. However, it is critical to highlight again Tamerlan's *six-month* trip to Dagestan in 2012.

Umarov repeatedly called for the move to a transnational jihad.[23] The Tsarnaev brothers are the first products of the Caucasus Emirate to take their fight global. New generations of Chechens are raised in a culture of martydom and animosity for Russians and enemies of Islam.[24] The dangers and risks for Western countries now are the radicalization in the Chechen diaspora. Unfortunately U.S. intelligence failed to recognize Tamerlan as a threat to the U.S. national security after warnings from the government of Russia. If the United States continues to underestimate the threat of the Caucasus Emirate, it will lead to increased vulnerability and potential future attacks.

The days of insurgency and separatist struggles in the North Caucasus are long gone. As a result of Russian oppression and Middle-Eastern jihadist-Salafist exploitation, the North Caucasus is a breeding ground for terrorists. By the early 2000s this much was obvious to the Russian government who inadequately and inhumanely tried to counter the imminent terrorist threat. However, the Caucasus Emirate is no longer a danger just to Russia; Western countries are at risk too. Any doubtful individual should visit *Kavkaz Center*. The global risk of the Caucasus Emirate and North Caucasus terrorism was proven true April 15, 2013 in Boston.

## Notes

1. Carlotta Gall and Thomas de Wall, *Chechnya: Calamity in the Caucasus* (New

York: New York University Press, 1998), 36.

2. Roland Dannreuther, "Islamic radicalization in Russia: an assessment," *International Affairs* Volume 86 Issue 1 (January 2010,) 112.

3. Yossef Bodansky, *Chechen Jihad: Al Qaeda's Training Ground and the Next Wave of Terror* (New York: HarperCollins Publishers, 2007), 3.

4. Michail Logvinov, "The Caucasus Emirate: Origins, Ideological Orientation, and Risk Status," *KAS International Reports,* March 2012, 88.

5. Bodansky, *Chechen Jihad,* 42.

6. Daniel Byman, *Deadly Connections: States that Sponsor Terrorism* (Cambridge: Cambridge University Press, 2005), 161.

7. David Satter, "The Russian Stake in Afghanistan," *Foreign Policy Research Institute*, March 2013. http://www.fpri.org/articles/2013/03/russian-stake-afghanistan

8. Mia Bloom, *Bombshell: Women and Terrorism* (Philadelphia, PA: University of Pennsylvania Press, 2011), 45.

9. Bloom, *Bombshell: Women and Terrorism,* 51.

10. Laura Sjoberg and Caron E. Gentry, *Mothers, Monsters, Whores: Women's Violence in Global Politics* (New York: Zed Books, 2007), 107.

11. Bodansky, *Chechen Jihad,* 288.

12. David Satter, *It Was a Long Time Ago, and it Never Happened Anyway: Russia and the Communist Past* (New Haven, CT: Yale University Press, 2012), 304.

13. Bodansky, *Chechen Jihad,* 282.

14. "Emir Seyfulla on preparation for the proclamation of the Caucasus Emirate" («Амир Сейфулла о процессе подготовки к провозглашению Кавказского Эмирата»), *Kavkaz Center,* last modified November 20, 2007, http://kavkazcenter.com/russ/content/2007/11/20/54479.shtml (accessed April 27, 2013).

15. "Russia: Ten Jailed for Nevsky Express Train Bomb," *British Broadcasting Corporation*, May 22, 2012, accessed April 27, 2013. http://www.bbc.co.uk/news/world-europe-18162643

16. "Al-Qaida Sanctions List," *United Nations,* July 3, 2012, accessed April 27, 2012.

http://www.un.org/sc/committees/1267/AQList.htm

17. Vladimir Bobrovnikov, "Al-Azhar and Shari'a Courts in Twentieth-Century Caucasus," *Middle Eastern Studies* Vol. 37, No. 4 (Oct 2001), 10.

18. "Who influenced the Tsarnaev brothers? Prof. Goble testifies on Islamist extremism in Chechnya," *Institute of World Politics*, posted April 26, 2013 http://www.iwp.edu/news_publications/detail/who-influenced-the-tsarnaev-brothers-prof-goble-testifies-on-islamist-extremism-in-chechnya.

19. http://www.youtube.com/user/muazseyfullah/videos?view=0

20. Muaz meaning distinguished or glorified in Arabic.

21. In 2012, Adam Lanza was responsible for the tragic shooting in Newton, Connecticut and James Holmes was responsible for Aurora, Colorado.

22. At this point in the investigation, little is known about the individual who provided guidance and support to the Tsarnaevs.

23. Logvinov, "The Caucasus Emirate", 88.

24. Bloom, *Bombshell: Women and Terrorism,* 67.

# Should High-Profile Terrorists and Cartel Leaders Be Eliminated? Targeted Killings in Latin America

W. Alejandro Sanchez

In September 2013, the Mexican newspaper *Proceso* reported that the U.S. government had plans to eliminate Joaquin "El Chapo" Guzman, the leader of the Sinaloa drug cartel, without notifying Mexican authorities.[1] The revelation, while explosive, did not appear to have damaged bilateral relations. Nevertheless, the prospect of an American military strike team entering Mexican territory to eliminate a high value target such as El Chapo Guzman is comparable to the May 2011 U.S. operation in Pakistan that eliminated Al Qaeda terrorist leader, Osama bin Laden.

The practice of targeted killings is controversial and complicated, not only for operational and legal reasons, but also from a moral standpoint. Arguably, there are a number of "bad guys" in the Western Hemisphere in addition to El Chapo Guzman who could qualify for this treatment, ranging from terrorist leaders to major cartel bosses. Regardless, determining which method is more effective, assassination or imprisonment of high-level criminals, requires a case-by-case study.

***W. Alejandro Sanchez*** *obtained a certificate in International Politics from IWP. He is a research fellow at the Council on Hemispheric Affairs, and a regular contributor to VOXXI News. His analyses have appeared in numerous journals including Small Wars and Insurgencies, Defence Studies, the Journal of Slavic Military Studies, European Security, Studies in Conflict and Terrorism, and Cuban Affairs.*

Most current analyses on targeted killings focuses on violent movements, such as al-Qaeda, the Taliban and Hezbollah. This paper analyzes target killing operations from a Latin American perspective due to of current Western Hemispheric security issues.

## Targeted Killings in the Western Hemisphere

It is inaccurate to consider terrorists, drug traffickers, transnational gangs, and insurgents as the same class of criminals, in spite of their similarly destructive *modus operandi*. The motivations, ideology, and objectives, as well as structures, of the numerous Latin American criminal entities are different. There are, however, common threads among organizations such as the Medellin cartel in Colombia, the Shining Path in Peru, and the Sinaloa cartel in Mexico, as all have become major security threats inside, and sometimes outside of, their respective nations.

As Latin American countries are increasingly confronted with violence and security threats, regional governments may resort more often to targeted killing operations instead of search-and-capture operations. Nevertheless, it is debatable if the elimination of a group's leadership may necessarily halt its operations permanently. For example, terrorist movements may react differently than drug cartels when a leader is eliminated. Furthermore, there are cases in which search-and-capture operations have been successful in affecting a criminal organization's structure, as was the case when Peruvian security forces captured major terrorist leaders.

**Eliminated Targets.** Among the leaders of violent organizations killed by targeted killing operations in recent years, Osama bin Laden's assassination stands out in the Western world, but Israel has also employed targeted killings to eliminate leaders of Hezbollah and Hamas.[2] Such operations are likely to continue, with the missions increasingly carried out by predator drones instead of Special Forces.[3]

Latin American governments, too, have resorted to targeted killings. Pablo Escobar, head of Colombia's infamous Medellin cartel, was shot to death in 1993 during a police raid.[4] Escobar rose to prominence in Colombia during the 1980s, becoming infamous for acts violence against drug rivals, civilians, and government officials. In one of his boldest operations, Escobar ordered the death of Colombian presidential candidate Luis Carlos Galan in 1989.[5]

Colombian security forces carried out a targeted killing operation against the leadership of the insurgent movement, the Revolutionary Armed Forces of Colombia (FARC) in 2008, bombing an Ecuadorian camp where members of FARC's leadership, including the group's then-leader Raul Reyes, were suspected of hiding.[6] The bombing raid killed Reyes and several other insurgents, but the backlash brought Colombia to the brink of war with Ecuador, which viewed the operation as a violation of its sovereignty. To further complicate matters, the Venezuelan government deployed its army to the Colombian border and Venezuelan President Hugo Chavez declared the possibility of going to war with Colombia to protect Ecuador's sovereignty.[7] In recent years, the Colombian military has been successful in eliminating other major FARC leaders: Jorge Briceno (aka Mono Jojoy) in September 2010, and Alfonso Cano in November 2011.

Similarly, the Mexican military has managed to eliminate the leadership of some of Mexico's major drug cartels. For example, Heriberto Lazcano Lazcano (*aka* El Lazca) of the Zetas Cartel, was shot to death by Mexican marines last October.[8]

**High-Profile Individuals Still Free.** In Latin America, numerous leaders of violent organizations remain at large and are likely subjects for targeted killings. Some examples of potential targets in Mexico include El Chapo Guzman, the ruthless head of one the most powerful and violent drug cartels in Mexico with significant wealth from drug trafficking;[9] Miguel Angel Trevino (aka Z-40), leader of the dreaded Zetas cartel; and Fernando Sanchez Arellano (aka The Engineer), head of the Arellano cartel.[10]

Leaders of hemispheric insurgent movements that could be targets for TK operations include Rodrigo Londono (aka Timochenko), current leader of the FARC; Nicolas Rodriguez Bautista (Gabino), the insurgent group's second in command; Jose Benito Cabrera (aka Fabian Ramirez),[11] current commander of the National Liberation Army (ELN); and the leader of Peru's Shining Path, Victor Quispe Palomino (aka Jose), who is believed to be operating in the Peruvian highlands.[12]

**High-Profile Individual In Prison.** On the other hand, various operations have led to the capture and imprisonment of high-profile targets in Latin America, including Abimael Guzman, the leader of Shining Path, and Victor Polay Campos, the head of the Tupac Amaru Revolutionary Movement (MRTA) in Peru. These criminals have been in maximum-security prisons since 1992, under life sentences. Other high-profile criminals include Salvador Alfonso Martinez (The Squirrel), a Zetas leader linked to more than 300 murders, one of which was an American citizen.[13] In Brazil, several leaders of the gang known as the First Capital Command (PCC) have been placed under arrest. These examples of successful imprisonments demonstrate that viable options exist that do not involve targeted killings.

## Analysis

Whether targeted killings are more effective than capture and imprisonment in removing high profile criminals from power depends largely on the capacity of a country's judicial and penal system to conduct fair trials of criminals and to keep them successfully in prison. Peru has been successful in keeping top terrorist leaders behind bars, but this hasn't always been the case. For example, the MRTA's leader, Polay Campos, was captured in 1989 and escaped from prison, along with 47 other MRTA members in 1990.[14] He was finally recaptured two years later. In addition, Pablo Escobar spent little over a year in a Colombian prison (1991–1992) before escaping. He remained on the run until he was shot in 1993.[15]

Meanwhile, the Mexican penal system's ability to handle a "big fish" such as El Chapo is questionable. There have been several recent jailbreaks in Mexico and dozens of cartel members have managed to escape.[16] Moreover, imprisoned criminal leaders are able to continue communicating with their subordinates outside of prison—the leadership of the Brazilian PCC, for example, has conducted conference calls from behind prison walls.[17] Hence, targeted killings for major criminal leaders may be the only option for some security forces, given the state of many Latin American penal systems that allow jailed criminals to continue giving orders to their henchmen.

The question remains as to whether targeted killing operations are successful in affecting a criminal organization's leadership structure. Osama bin Laden's death crippled al-Qaeda , and Escobar's death dissolved the Medellin cartel. On the other hand, imprisonment of Abimael Guzman and Polay Campos was enough to severely cripple the Shining Path and the MRTA respectively.

As previously mentioned, a critical factor is the leadership structure of criminal movements. FARC's leadership, unlike that of the Shining Path, does not revolve around one individual. Rather, a secretariat selects a new leader when one is eliminated,[18] thereby enabling the group to withstand the loss of major leaders. Furthermore, the structure and lack of political and ideological objectives of drug trafficking groups, such as Mexico's cartels, make it likely that the Sinaloa cartel could survive if El Chapo Guzman was eliminated. The result would likely be a factionalized Sinaloa cartel and continued drug trafficking operations. A September 14, 2013, blog post in the *Houston Chronicle* argues that cartels such as the Gulf Cartel and Felix Arellano Organization have survived the arrest of their major leaders by aligning themselves with stronger cartels (such as the Sinaloa or the Zetas). In addition, the Gulf Cartel "has deep connections in the social fabric of northeastern Mexico [and] it boasts of extensive U.S. domestic wholesale transportation networks."[19] With such a history and network, this cartel has survived the fall of its major bosses over the past decade, demonstrating that targeted killings do not always have the intended effects.

## Conclusions

Some experts who have written on targeted killing operations against groups such as al-Qaeda argue that Washington and Tel-Aviv have set "a dangerous precedent for abusive regimes around the globe to conduct drone attacks or other strikes against persons who they describe in vague or overly broad terms as terrorists."[20] In Latin America, this precedent has arguably already been set with the 1993 raid on Pablo Escobar and the 2008 Colombian military raid in Ecuador to eliminate a FARC leader.

In sum, it is not certain that targeted killings have had, or ever can have, a crippling effect on criminal entities in the Western Hemisphere. Arguably, the Sinaloa Federation could survive the elimination of its leader, El Chapo Guzman. Analysts have argued "the Federation is just that—a loosely knit alliance of smaller cartels that owe allegiance to El Chapo and his people. When a kingpin is captured or killed, such alliances tend to fracture as it becomes a case of every man for himself."[21] In other words, while the elimination of El Chapo might weaken or even break apart the Sinaloa cartel, it would probably just lead to a factionalized organization, with little to no effect on drug trafficking itself.

Criminal entities in the Western Hemisphere range from narco-terrorist groups in Peru and Colombia, to expanding drug cartels in Mexico, and powerful gangs in Brazil and Central America. Each entity has its own *raison d'etre* and leadership structure, which means that some could arguably better withstand the fall of its leader more than others (for instance, the Sinaloa cartel compared to the Shining Path). Thus, specific analyses of each organization are essential to better understand whether a targeted killing of its leader would be successful. So far, the effectiveness of targeted killing operations, as compared to search-and-capture operations, has yet to be proven in the Western Hemisphere.

## Notes

1. Jorge Carrasco and Jesus Esquivel, "Mision del Pentagono: Atrapar a 'El Chapo'.... O acabar con el," Proceso (Mexico), August 11, 2012. Accessed December 10, 2012: http://www.proceso.com.mx/?p=316815

2 "Israel's Targeted Assassinations: An Overview Of Militants Killed By Israel (PHOTOS),"

*The Huffington Post*, November 14, 2012. Accessed December 22 http://www.huffingtonpost.com/2012/11/14/israel-targeteded-assassinations_n_2131591.html . Also see: Gal Luft, "The Logic of Israel's Targeted Killing," *The Middle East Quarterly*, Winter 2003, Volume X, No. 1, Winter 2003, p. 3-13. Accessed December 22, 2012, http://www.meforum.org/515/the-logic-of-israels-targeteded-killing

3 Joshua Fost, "Targeted Killing, Pro and Con: What to make of U.S. Drones in Pakistan," *The Atlantic*, September 26, 2012. Accessed December 18, 2012 http://www.theatlantic.com/international/archive/2012/09/targeted-killing-pro-and-con-what-to-make-of-us-drone-strikes-in-pakistan/262862/ Also see: Akbar Nasir Khan, "The US' Policy of Targeted Killings by Drones in Pakistan," *IPRI Journal*, XI, no 1, (Winter 2011), p. 21-40

4 Mark Bowden, *Killing Pablo*, (New York: Penguin Books, 2002).

5 "El sacrificio y la impunidad en el caso Galan," *El Espectador* (Colombia), September 13, 2012. Accessed December 22, 2012 http://www.elespectador.com/especiales/articulo-374565-el-sacrificio-y-impunidad-el-caso-galan

6 "La muerte de 'Raul Reyes' desencadena una crisis diplomatica entre Colombia, Venezuela y Ecuador," *El Pais* (Spain), Internacional, March 2, 2008. Accessed December 21, 2012 http://internacional.elpais.com/internacional/2008/03/02/actualidad/1204412408_850215.html

7 Ian James, The Associated Press, " Chavez Deploys Troops to border with Colombia," *The Spokesman Review*, March 3, 2008. Accessed December 21, 2012 http://www.spokesman.com/stories/2008/mar/03/chavez-deploys-troops-to-border-with-colombia/

8 Silvia Otero and Doris Gomora, "Confirman la muerte de Lazcano, lider de 'Zetas,'" *El Universal* (Mexico), October 10, 2012. Accessed December 17, 2012 http://www.eluniversal.com.mx/notas/875733.html

9 "NARCO BLOG: Sinaloa Cartel Welcomes Police Chief with Tortured Body," *El Blog del Narco, Hispanically Speaking News*, March 12, 2011. Accessed Decembe20, 2012 http://www.hispanicallyspeakingnews.com/el-blog-del-narco/details/narco-blog-sinaloa-cartel-welcomes-new-police-chief-with-tortured-body/6033/ . Regarding Guzman's known wealth see: Erin Carlyle, "Billionaire Druglords: El Chapo Guzman, Pablo Escobar, The Ochoa Brothers," *Forbes.com*, March 3 2012. Accessed January 4, 2013 http://www.forbes.com/sites/erincarlyle/2012/03/13/billionaire-druglords-el-chapo-guzman-pablo-escobar-the-ochoa-brothers/

10 For other cartel bosses see: "De Siete al Inicio del Sexenio, los carteles mexicanos se multiplicaron: Calderon dejara por lo menos 25," *Tiempo Real* (Mexico), August 20, 2012. Accessed January 4, 2013 http://www.sinembargo.mx/20-08-2012/335244

11 "Reaparece importante jefe de las FARC dado por muerto en Colombia," *Univision*, July 31, 2012. Accessed December 16, 2012 http://noticias.univision.com/america-latina/colombia/article/2012-07-31/reaparece-jefe-farc-dado-por-muerto-television-colombia#axzz2Fw5cCurP

12 Hans Huerto Amado, "Quien es el proximo cabecilla de Sendero Luminoso a ser neutralizado?" *El Comercio* (Peru), February 13, 2012. Accessed December 18, 2012 http://elcomercio.pe/politica/1373857/noticia-quien-proximo-cabecilla-sendero-luminoso-neutralizado

13 Allison Jackson, "Mexico arrests Los Zetas drug cartel leader 'The Squirrel' over migrant massacres, US tourist murder," *The Global Post*, October 8, 2012. Accessed December 20, 2012 http://www.globalpost.com/dispatch/news/regions/americas/mexico/121008/mexico-los-zetas-drug-cartel-the-squirrel-migrant-massacres-murder-US-tourist

14 Ana Murillo, "Detenido en Peru el jefe del Tupac Amaru," *El Pais* (Spain), February 5, 1989. Accessed December 12, 2012 http://elpais.com/diario/1989/02/05/internacional/602636409_850215.html . Also see: Carlos Castillo, "A 20 anos de la fuga de Victor Polay," *El Comercio* (Peru), May 31, 2010. Accessed December 12, 2012 http://peru21.pe/noticia/487763/20-anos-fuga-polay

15 See Bowden's book for in-depth details of Escobar's lavish lifestyle behind bars.

16 Oscar Villalba, "Mexico Prison Break: More than 130 Inmates Escape from facility near U.S. border," *Huffington Post*, September 17, 2012. Accessed December 15, 2012 http://www.huffingtonpost.com/2012/09/17/mexico-prison-break_n_1892209.html . Mexico also has problems in its judicial system. See: Patrick Corcoran. "Corruption could be undoing of Mexico's judicial reforms," *MexicoData.Info*, March 17, 2008. Accessed January 4, 2013: http://mexidata.info/id1754.html

17 Jack Davis, "Brazil Prison Gang conducted 10-Hour Conference Call," *InSightCrime*, December 5, 2012. Accessed December 22, 2012 http://www.insightcrime.org/news-briefs/brazil-prison-gang-conducts-10-hour-conference-call

18 "Colombia Peace at last?" International Crisis Group, Latin America, Report N. 45, September 25, 2012, p.13.

19 Nathan Jones, "Gulf Cartel will likely survive arrest of high-level leaders," *Houston Chronicle*, September 14, 2012. Accessed, December 23, 2012. http://blog.chron.com/baker-blog/2012/09/gulf-cartel-will-likely-survive-arrest-of-high-level-leaders/

20 "Letter to Obama on Targeted Killings and Drones,"*Human Rights Watch*, December 7, 2012. Accessed December 21, 2012 http://www.hrw.org/news/2010/12/07/letter-obama-targeted-killings

21 Sylvia Longmire, "Why arresting 'El Chapo' might be a bad thing," *Small Wars Journal*, October 31, 2012. Accessed December 21, 2012 http://smallwarsjournal.com/jrnl/art/why-arresting-%E2%80%9Cel-chapo%E2%80%9D-might-be-a-bad-thing

# Subversion of American Higher Education

Nicholas Milan

Twenty-one years after the implosion of the Soviet Union, the subversive seeds planted by the Communists to create social upheaval in American society continue to grow, namely in America's educational system. These subversive activities are rooted in earlier events of the 20th century. The violence of the previous century was led by competing ideologies—fascism, communism, socialism, and capitalism—which lead to "more human beings…slaughtered in the 20th century than all previous centuries combined."[1] The consequences of such destruction and disorder influenced the geopolitical composition of the international system, ultimately dividing the world into two camps. Communism and its stronghold, the Soviet Union, was juxtaposed to democracy and capitalism and its defender, the United States. Communists in many countries saw their entrance into the political domain through revolution, seizing power through manipulating populations with propaganda, disinformation, and political warfare. As the "red tide" spread to countries, it was met with opposition by the United States and its allies. Communist leadership was unsure of how to eliminate the staunch opposition of the United States. Communists sought to answer a question posed by the French Count Maistre a few centuries earlier: "[N]ations were killed by conquest, that is by invasion: but here an important question arises; can a nation not die on its own soul, without resettlement or invasion, by allowing the files of decompo-

***Nicholas Milan*** *is currently enrolled in the National Security and Statecraft program with a specialization in Intelligence, he will graduate in 2013. Mr. Milan graduated from Saint Joseph's University in Philadelphia with a BA in International Relations. He is a former high school teacher.*

sition to corrupt to the very core those original and constituent principles which make it what it is?"[2]

Communists engaged the U.S. in psychological and political warfare. As Lenin foresaw, "First we will take Eastern Europe; then the masses of Asia. Then we will encircle the United States of America which will be the last bastion of Capitalism. We will not have to attack it; it will fall like an overripe fruit into our hands."[3] The objective was to subvert the American moral fabric based upon Judeo-Christian values, to create a fertile ground for revolution in the U.S., in which the communists would seize power. Yuri Bezmenov, a defected KGB officer, gives insight into the Soviet Union's operations:

> We rarely use guns to kill people and take their country. The cleanest way is to blackmail, pervert, bribe, lie and intimidate the POLITICIANS and the MEDIA, and they will destabilize and disunify their own country for us. Then all we have left to do is to arm the procommunist or simply criminal factions and we have a coup and another "liberated" country. As neat as that.[4]

Bezmenov's explanation of Soviet methodology is further supported by the book *Peaceful Coexistence; a Communist Blueprint for Victory:* "The communists would prefer to come to power by means of the peaceful method...their willingness to utilize the instruments of democracy to destroy that very democracy which guarantees them the right to organize politically and even take part in the conduct of governmental affairs."[5] In achieving this goal, there were several strategies employed to erode American opposition to communism. America knew that ideological war was ensuing and dividing the world, but it did not know the battlegrounds which it would engage, "a war about which the average American hasn't the foggiest idea. The reason Americans do not understand this war is because it has been fought in secret—in the schools of our nation, targeting our children who are captive in classrooms."[6] American education became a prominent target for communist subversive activity, as reflected in the conclusions by the Subcommittee to Investigate the Administration of the Internal Security Act and other Internal Security Laws: "World Communist leaders have made schools and colleges of the United States a target of infiltration and activity as part of their program to destroy the United States."[7] The ideological war waged by communists upon the United States has lasting currents that are still rippling in society today.

## The Focus of Subversion

The attractiveness of American education rested in the strategic importance of the communist goal and recognition that the current generation was inherently opposed to communism because of education in a capitalist, democratic society. John Dewey, a proponent of the Soviet Russian education system, remarked upon this problem that "the educator found that the work he was trying to do in school...was undone...from the environment."[8] The environment referenced by Dewey is family, church, and other moral, nationalistic influences in the child's maturing. In 1924, Stalin recognized the importance of education and culture manipulation stating cultural and educational organizations are valuable allies in the communist battle for world dictatorship.[9] The communist, determined to rise to power, that fertility for revolution was necessary, which made it necessary to decay the fabric of American society, which required the decay of inherent resistance to communism, and this entailed indoctrinating American youth to be open to change and communist ideals: "Communists look to create the conditions that are favorable to a revolution within a country"[10] and "the best way to do this is to target the youth, the earlier ages, to indoctrinate them and when they grow, they will be opened to such ideals or be the leaders of it."[11] The logic of attacking the education system is reflective of the tactics used by the Bolsheviks when they seized power in 1917, destroying the old educational system and instilling a new education curriculum.[12] Lenin so strongly believed in the communist methodology of indoctrination that he stated, "Give us the child for eight years and it will be a Bolshevik forever."[13]

Many Americans did not understand the location and target of the Soviets, they believed it was a governmental, national conflict, yet it was more personal. The ultimate target of the communist strategy was the minds of the American people, as reflected by Yuri Bezmenov, "the war is not against the capitalist system, but against the individual American."[14] The Soviet Union was unequivocally committed to ideological subversion and the KGB appropriated up to 85 percent of its budget to active measures.[15] The appropriation of resources reflects the focus of the "master-subverters"[16] on youth, because of their malleability and gullibility, as a direct order from Moscow: "The university where young men and women gather and discuss... is the ideal seeding ground for revolutionary ideas. New ideas take root swiftly and flower abundantly."[17] Centralized education provided a top-down approach in controlling education and its curriculum, therefore constructing a homorganic approach in creating a new society. Although the Soviet Union strategically subverted the education system intensively during the Cold War, subversion began after the Bolshevik Revolution and was highlighted by a concentrated small group of Marxists whose members began disseminating their ideology in 1935 in New York City.

## The Frankfurt Institute of Social Research

The Frankfurt Institute of Social Research (Frankfurt School) relocated from Germany to New York after the rise of Adolf Hitler. The Frankfurt School was a Marxist-grounded institution, and provides a synoptic insight into the Soviet Union's ideals and planned subversion of the United States. Essentially:

> The Frankfurt School believed that as long as an individual has the belief that his divine gift of reason could solve the problem facing society, then that society would never reach the state of hopelessness and alienation that they considered necessary to provoke socialist revolution. Their task, therefore, was as swiftly as possible to undermine the Judeo-Christian legacy.[18]

Soon the Frankfurt School became affiliated with Columbia University while a diaspora of its intellectuals settled in universities such as Princeton, Brandies, and California at Berkley. The professors' teachings, biased toward Marxist ideology, would prove venomous for American society. The objective of the Frankfurt School and its successors is to control the U.S. educational systems and produce poorly educated, little-world citizens, to forego the liberties of the Constitution.[19]

These methods and the rhetoric of these ideals and goals would evolve through the century and remerged in a communist meeting at the University of California, Berkley in 1992. Akin to the Soviet Union, the Frankfurt School understood the importance of teachers as agents of change. It was through these agents of change, placed in selective positions of influence and authority, that communists would bring about transformation in American society. Teachers, educators, journalists, politicians, all held prominent status in society and therefore their words and deeds would have increasing influence over the population. Willi Munzenberg, a German communist and political activist, understood their usefulness, stating, "Organize the intellectuals and use them to make Western civilization stink. Only then, after they have corrupted all its values and make life impossible, can we impose the dictatorship of the proletariat."[20] Indoctrinating a new generation of Americans through institutions of influence would produce change; however this change would require patience to achieve. The purpose of gradual change is to avoid shocking the population with radical ideas and behavior, inciting a strong negative response, but to slowly and incrementally advance toward communist views, without the population consciously knowing and recognizing the direction of change. One method that Soviets and "useful idiots" used to integrate their ideology was semantic deception, "creating or using words that are affiliated with specific definition or values but are actually used for alternative purposes and agendas."[21] The use of semantic deception would entrap parents and leaders to act in a specific manner to support their children's education, while it was actually hindering it. The Frankfurt School exemplifies this point, by avoiding the terms "Marxism" and "communism," and replacing them with "materialist theory of society" to identify their subject matter. Eventually, through these two methods, the population will not perceive the enemy as an enemy, but view the enemy as a friendly alternative to the current system.

## Foundation of Progressive Education

"Progressive education" was founded by American socialists at the turn of the 20th century, who, because of their socialist ideals, proved an easy target for communists to infiltrate. The Progressive Education Association (PEA) was founded in 1919 by "useful idiot" John Dewey.[22] The PEA believed that education should be based on "progressive science," and that it should instill the ideology of communism in the minds of young generations. The disciples of progressive education must also be brought to the realization that the doom of the capitalistic world is inevitable and therefore must actively prepare for a new world, by embracing change.[23] Progressive education, as infiltrated by communists, centered on social collectivism, and sought to disintegrate human loyalties and objective values: "[E]ducation should shape students, to survive such an environment, who are not bound by deep loyalties, hold all conclusions and values tentatively and be ready to make fundamental shifts in outlook and philosophy."[24] Again, this is a reflection of the communist objective of subverting America, and its taking root in American education in the 1930s. In validation of the PEA and its leadership's intentions, PEA member O.A. Nelson attended a PEA meeting in 1928 with Dewey. The topic of the meeting was the subject of math and how to "better" teach students. Listening to the speaker, Nelson was informed that the purpose of math was to dumb down students. Nelson's rejection of this idea caused him to leave the PEA. He later admitted, "I thought the word 'progressive' meant progress for better schools," validating the semantic deception of the communists. Nelson also stated that, "Eleven of those attending the meeting were leaders in education, Drs. John Dewey and Edward Thorndike, from Columba University were there and the others were of equal rank. I checked later and found that all were paid members of the Communist Party of Russia."[25] The progressive movement's infiltration was led by "useful idiots,"[26] achieving the goals of Soviet demoralization, as expressed by William Z. Foster, the head of the Communist party USA: "Our teachers must write new school textbooks and rewrite history from the Marxian viewpoint."

## Useful Idiots

The Soviets used sympathetic American political and social organizations to amplify their influence. This allowed the communists to change the appearance of their identity in the West so they wouldn't automatically be disregarded; so they joined and supported liberal parties and organizations. These organizations and their members provided a base and face for the communists to gain popular support.[27] One example was the utilization of the progressive movement in America to gradually introduce communist policy and ideology into society and education. As the two became further intertwined, the progressive movement supported the idea that schools should be "seminaries for indoctrination."[28] Secondly, it proved difficult for American citizens to oppose an organization and movement that was partially focused on education and named "progressive." Opposition to progressive ideals meant that one was opposed to progress, opposed to better public education. The communists assisted the change of their identity by naming their centers after historical American figures, such as the Thomas Jefferson School, the John Adams School, and the George Washington School.[29] Further along, the hijacking of the progressive movement and organizations, individuals emerged to spearhead the subversion of American education. Some of these are John Dewey, George Counts, Antony G. Oettinger, Howard Zinn, Dr. George Hartmann, and William Z. Foster. Lenin knew the usefulness of sympathizers to communist ideology and referred to them as "useful idiots." These useful idiots would prove to be the gateway into American society and, especially, into the education system.

## Techniques

According to the Soviets, ideological subversion is a long-term process that involves four sequential stages of varying lengths: (1) demoralization, (2) destabilization, (3) crisis, and (4) normalization. Demoralization is the primary step in subverting a nation and requires the most time. "The purpose of demoralization is to change the perception of reality of every American to such an extent that despite the abundance of information, no one is able to come to sensible conclusions in the interest of defending themselves, their families, their community and their country."[30] It would render a significant portion of the citizenry vulnerable to Marxism ideology, and unclear about the subverters' true intentions. The benefit of demoralization is that the targeted population is unaware of being demoralized, through gradualism, and allows the enemy to pursue its goals without the people being aware.[31] Demoralization focuses on the sector of society that retains influence, such as, religion, the education system, social life, administration, and the military. Paralleling the subversion of influence outlets, communists also sought to incite and support opposition groups, such as criminals, political minorities, and radical movements. The educational system and universities provided numerous opportunities for subversion, through the classroom, student organizations and social life, to engage the students in protests and infect their minds with communist ideology.

Teachers and professors received several guides and instructions on classroom techniques and strategy. A basic technique, shared by "useful idiots" and communist fronts was to avoid being labeled as communist. This principle was transferred to the agents of change in schools. Educators should not be explicit in their views, teachings and orientation. They are to carefully choose opportunities to inject communist lessons and ideology without students realizing what they are being taught.[32] Specific tactics that professors employed to achieve their ends included recommended reading lists on syllabi, selecting literature written by authorities with pro-Soviet positions, librarians stacking bookshelves with pro-communist literature, and infiltrating teacher groups and parent associations to adjust the curriculum. One example is *A People's History of the United States* by Howard Zinn, which was nominated for the National Book Award in 1980. The book approaches the history of the U.S. from a societal point of view of a quiet revolution through the eyes of the common person and not the elites of society. The culmination of the teachers' activities is to ideologically soften the students to communist ideology. Outside of the individual professors, other educational activities were planned to encircle the students' life with hidden communist ideology. Some of these activities included student exchanges whereby American students and professors travel to Moscow and are exposed to an ideological barrage, saturation of campus bookstores with Marxist and socialist literature, inviting American students and professors to international conferences with Soviet participation and Soviet propaganda, establishment of numerous student newspapers and magazines organized by communists and "useful idiots," and organizing study groups for dissemination of Soviet propaganda and communist ideology.[33] Additional techniques included arranging a celebrity, musician, and person of notoriety to campus and disseminate propaganda, as well as manipulating students by "espousing popular causes and protesting militantly against anything which they can make appear as unfair practice, exploitation or discrimination."[34] Professor John Hanna of Columbia University testified that communists can manipulate admissions and scholarships to recruit radical students and mandate courses that maximally expose students to indoctrination.[35] The author of *You Can Trust the Communists: to Be Communists,* Charles Schwarz, gives insight into the reasoning and manipulation of university students:

> The communists take advantage of such students and individuals at this point in their life. Students far from home, whether citizen or not are susceptible to foreign influence... The student is usually homesick and lonely. They make him feel important and interested in him. Provide him a social life and companionship. When they have won him, they equip him with organizing skills and the necessary tools to serve Communism in his homeland.[36]

The Communist Party in the West was open and instigated more distractions so that the master-subverters and their networks would receive less attention and therefore, work easier.[37] Fermenting distractions was not lost upon the agents of influence in the education system. The agents of influence would incite students to sow discord and conflict with institutions and the government. One example was the inciting of university students during the Vietnam War to protest the government and the armed forces, and to resent authority. Such actions received praise and notoriety from the Kremlin. The Kremlin publication *The Party Bulletin* praised the American subverters and their undercover networks for, "excellent direction of the U.S. students throughout America, which resulted in nationwide revolutionary actions."[38]

The progressive movement later adopted the Whole Language Method of Education, founded by Lev Vygotsky, a Marxist psychologist, which later evolved into the "balanced approach." The conclusion by some, is that the purpose of the Whole Language Method is to intentionally dumb down the students, as witnessed by O.A. Nelson. This served a dual purpose of first, reducing the intelligence of students, the future leaders, so they could not defend the subversion of America; and second of creating an artificial crisis, a distraction, as explained by Yuri Besmenov, to the rest of the demoralization process. According to Samuel Blumenfeld, Whole Language Method is founded upon detrimental educational principles. A few principles are: the process of not deriving meaning from text, but constructing one's own interpretation from the student's social structure to understand the text; and unrecognizing of an absolute, but only plausible, answer because answers are individual. This is the educational framework for the introduction of relativism into the classroom: values clarification.

## Success of Soviets

The patient persistence of the Soviets in engaging in ideological subversion led to early success, which continued beyond the Cold War. Demoralization had a distinct advantage over conventional warfare, and the Soviets exploited this advantage. Unlike a military invasion, where soldiers and weapons are indefinitely needed to occupy the territory, demoralization required just an injection of communists, and Marxist ideology, into society, and it was self perpetuating. Once a professor was infected, his students became infected, growing exponentially, without much societal awareness. A student protest in Japan in 1960 demonstrated the level of manipulability by communist subverters in organizing and influencing students. President Eisenhower cancelled his scheduled visit to Japan that year. J. Edgar Hoover recognized the success and manipulation of the Soviets over students and commented that,

> The successful communist exploitation and manipulation of youth and student groups throughout the world today...a major challenge which free world forces must meet and defeat. Recent world events clearly reveal that world communism has launched a massive campaign to capture and maneuver youth and student groups. Communists have become experts at using this force to create chaos.[39]

While the Soviets were concentrating on raising a new generation, either open to communism or too ignorant to defend America, they sought to implant their educational subversion in international organizations, to reinforce the subversion in America. The United Nations and its several organizations became a target. This became evident when some of the U.N. personnel started reiterating similar, if not identical, goals and ideals that the communists outlined earlier in the century for subverting America. One example is Julian Huxley, the first director of the United Nations Educational, Scientific and Cultural Organization (UNESCO), from 1946 to 1948. He was a communist that held an influential position in organizing global education to have similar standards and goals for students around the world. He assisted in the composition of the *Handbook of Teachers*, which described one of the main functions of teachers as preparing students for change. In regard to people's identity, they should be ashamed of their nationality, and UNESCO is to promote world citizenship. The handbook went further to identify the necessity of schools, influencing children from youth, "before the child has entered school, his mind has already been profoundly marked and often injuriously by earlier influences."[40] These earlier influences refer to the parents and church, as earlier identified by Dewey. The teachings and guidance of parents are viewed as errors, because they are still educated under the old system. This conclusion is supported by a more explicit statement by Huxley that, "it is frequently the family that infects the child with extreme nationalism. The school should therefore... combat family attitudes."[41]

Huxley is validating an active approach in demoralization by decaying the family unit and having the state supersede the family's place and purpose. These subversive ideas are continued in later UNESCO publications and educational material. One UNESCO publication, *In the Classroom with Children Under Thirteen Years of Age*, states that methods are needed to "correct many of the errors of home training which cultivate attitudes running directly counter to the development of international understanding... As long as the child breathes the poisoned air of nationalism, education in world-mindedness can produce only precarious results."[42]

The Subversive Influence in the Educational Process hearings before the 82nd Congress Subcommittee to Investigate the Administration of the Internal Security Act and other Internal Security Laws of the Committee on the Judiciary United States Senate in 1953, provides further insight into the success and location of communists in education. Dr. Bella Dodd, a former member of the National Committee of the Communist Party and former professor of Hunter College testified to several accounts of infiltration. When Dr. Dodd was the leader of the New York Teacher's Union, 1,000 of the 11,000 members were also members of the Communist Party. She also recounted that the purpose of professors was to make themselves "an agent of class struggle and indoctrinate fellow professors"[43] while harnessing and directing the energy of the students.[44]

Charlotte Iserbyt, Senior Policy Advisor in the Office of Educational Research and Improvement (OERI) at the U.S. Department of Education under President Ronald Reagan, asserted that "there are people, who have been placed in the education system who seek to destroy absolute values and cast aside the importance of right and wrong that form the basis of any stable and free society."[45] Iserbyt is referencing the agents of influence, subverters, and their disciples, who have sought to undermine the education system by indoctrinating students with communist and collectivist ideals, and train humans like animals, not with real education. Iserbyt's comments indicate the success and continuation of educational subversion that started over 60 years prior, and the consequential effects they have had on education and society, the progress of relativism (values clarification), reduction of the church in society to decay morality, and the increasing acceptance of communism not as an enemy but just as an alternative economic theory.

A modern example of the continuation of communist strategic subversion is the account by former Idaho State Representative Curtis Bowers in the film *Grinding Down America*. In 1992, Bowers attended a Communist Party meeting at California Berkley to assess the American communist plans for life beyond the USSR. While attending the meeting, Bowers was astonished to find that the agenda proposed by the present communists, was reflective of an agenda set decades earlier by the Frankfurt School. Some of the goals they outlined to subvert American culture included the promotion of cohabitation over marriage, the support of the feminist movement to encourage women to neglect traditional roles, manipulation of the environmental movement to discourage business growth by excess regulation, and the acceptance of homosexuality in society.[46]

## Conclusion

The Soviet Union and its sympathizers have done their best to answer Count Maistre's question. The subversion of American education is no longer executed by the Soviet Union, however its sympathizers and subverters have been positioned to succeed after the fall of the Soviet Union. The Soviets understood that the strength of the United States was more than its military, but included its moral fabric, a reflection of its citizenry. Therefore, the USSR would have to attack the U.S., not by invasion, but with subversion from within, corrupting and decaying the individual citizen to not understand objective truth, not identify with his nation, and become softened to communist ideology. The book *The New Frontier of War* succinctly summarizes the Soviet value of subverters and "useful idiots":

> The Communists are aware of the fact that somebody who sympathizes with them is generally more valuable than a dozen militant communists. A professor who—without being a Communist Party member—stands up for the interests of the Soviet Union is more valuable than a hundred men with a membership book[47]

The ultimate ambition of the Soviet Union was to subvert America's education system to produce Soviet allies from within. The undermining and infiltration, without recognition, that America's future generations would be socialist and at best communist, "without his ever knowing it, a young American is thereby trained to be a potential Red ally."[48] The beginnings of the progressive movement and the arrival of the Frankfurt School snaked their entrance into positions of leadership and influence, publishing books, articles and educational materials to better serve the outlined goals of the communists after the Bolshevik Revolution. The effects are apparent throughout the last century as the disciples of the original subverters still carry out the work and mission of their teachers. Author Michael Minnicino observed in 1992 that, "The heirs of the first professors, infiltrated many universities, teaching their kids the ideology of the Frankfurt School, teaching their students to replace reason and teaching tolerance for movement from the left, but intolerance for movements from the right."[49] Demoralization has not succeeded in its absolute decay of American society but that is the purpose and danger of gradualism. It is critical to understand that today's youth are the government of tomorrow and, as the KGB defector Yuri Bezmenov stated, the eventual result is very predictable—ignorance combined with anti-Americanism.[50]

## Notes

1 *Agenda: Grinding America Down*, directed by Curtis Powers, Black Hat Films, 2010.

2 Timothy Matthews, "The Frankfurt School: Conspiracy to Corrupt," *Catholic Insight*, March 2009, http://www.scribd.com/doc/15584659/The-Frankfurt-School-Conspiracy-to-Corrupt-2009 (accessed October 2, 2012).

3 W. Cleon Skousen, *The Naked Communist*. (Salt Lake City: Ensign Pub., 1958), p. 65.

4 Tomas D. Schuman, *Love Letter to America* (Los Angeles: Almanac Panorama, 1984), p. 4.

5 American Bar Association, *Peaceful Coexistence; a Communist Blueprint for Victory*. (Chicago: American Bar Association, 1964), p. 42.

6 Charlotte Thomson Iserbyt, *The Deliberate Dumbing down of America: A Chronological Paper Trail* (Ravenna, OH: Conscience Press, 1999), p. xvii.

7 S. Rep. No. 83-76925-0 at 1 (1953).

8 John Dewey and William W. Brickman, *Impressions of Soviet Russia and the Revolutionary World: Mexico - China - Turkey, 1929* (N.Y.: Bureau of Publications, 1964), p. 70.

9 Louis F. Budenz, *The Techniques of Communism* (New York: Arno Press, 1977), p. 131.

10 American Bar Association, *Peaceful Coexistence; a Communist Blueprint for Victory*, p. 29.

11 *Agenda: Grinding America Down*.

12 U.S. House Committee on Un-American Activities, *100 Things You Should Know about Communism and Education*, series 3 (Washington D.C.: U.S. Government Printing Office, 1948), p. 4.

13 Ibid.

14 Tomas D. Schuman, *Love Letter to America*, p. 3.

15 "Psychological Warfare Subversion & Control of Western Society," interview with Yuri Bezmenov, February 2, 2011, http://www.youtube.com/watch?v=5gnpCqsXE8g&feature=related.

16 Joseph Bernard. Hutton, *The Subverters* (New Rochelle, NY: Arlington House, 1972), p. 97.

17 Ibid.

18 Matthews, Timothy, "The Frankfurt School: Conspiracy to Corrupt."

19 Alan Caruba, "Subversion of Education in America," Subversion of Education in America, 2001, http://www.deliberatedumbingdown.com/pages/reviews4.html (accessed October 1, 2012),

20 Ibid..

21 Charlotte Thomson Iserbyt, *The Deliberate Dumbing down of America: A Chronological Paper Trail*, p. 8.

22 Ibid., p. 7.

[23] Joseph Bernard. Hutton, *The Subverters*, p. 254.

[24] George S. Counts, *Dare the School Build a New Social Order?* (New York: Arno Press, 1969), p. 27.

[25] Charlotte Thomson Iserbyt, *The Deliberate Dumbing down of America: A Chronological Paper Trail*, p. 13.

[26] "Psychological Warfare Subversion & Control of Western Society," interview with Yuri Bezmenov, February 2, 2011, http://www.youtube.com/watch?v=5gnpCqsXE8g&feature=related.

[27] Ronald A. Joel, *The Strategy and Tactics of Communist Aggression*, 1st ed. (Alpha Printing, 1971).

[28] "Socialist Revolution in Public Schools,"Vimeo, http://vimeo.com/22263361 (accessed October 14, 2012),

[29] Joseph Bernard. Hutton, *The Subverters*, p. 74.

[30] "Soviet Subversion of the Free World Press." Interview by Yuri Bezmenov. April 24, 2011. http://www.youtube.com/watch?v=Cnf0I2dQ0i0.

[31] Ibid.

[32] Louis F. Budenz, *The Techniques of Communism.*

[33] Tomas D. Schuman, *Love Letter to America*, p. 7.

[34] U.S. Committee on Un-American Activities, *100 Things You Should Know about Communism and Education*, series 3 (Washington D.C.: U.S. Government Printing Office, 1948).

[35] Ibid.

[36] Schwarz, Frederick Charles, *You Can Trust the Communists: (to Be Communists)*, 7th ed. (Long Beach, CA: Christian Anti Communism Crusade, 1972).

[37] Joseph Bernard Hutton, *The Subverters*, p. 83.

[38] Ibid., p.103.

[39] William Roscoe Kintner and Joseph Zack Kornfeder, *The New Frontier of War*, p. 129.

40 Julian Huxley, *UNESCO: Its Purpose and Its Philosophy.* ([Washington]: Public Affairs Press, 1947).

[41] Ibid.

[42] Charlotte Thomson Iserbyt, *The Deliberate Dumbing down of America: A Chronological Paper Trail*, pg 345.

[43] S. Rep. No. 83-76925-0 at 1 (1953).

[44] Ibid.

[45] Charlotte Thomson. Iserbyt, *The Deliberate Dumbing down of America: A Chronological Paper Trail*, pg. 382.

[46] *Agenda: Grinding America Down*, dir. Curtis Powers.

[47] William Roscoe. Kintner and Joseph Zack. Kornfeder, *The New Frontier of War*, pg. 80.

[48] Skousen, W. Cleon. *The Naked Communist.* pg 78.

[49] Charlotte Thomson Iserbyt, *The Deliberate Dumbing down of America: A Chronological Paper Trail*, p. 389.

[50] Tomas D. Schuman, *Love Letter to America*, pg.8.

# The Meaning of the Shebaa Farms Dispute

Samantha Borzi-Hedges

Throughout modern history, the Shebaa Farms area has been contested as to whether it belongs to Syria or Lebanon. This conflict was mostly inconsequential until the Lebanese militant organization Hezbollah started staging attacks in this area propelling it into international headlines and making the resolution of the conflict imperative to regional security. Shebaa Farms is of no great geographic importance, but has become politically important as a staging ground for the ongoing conflict between Lebanon and Israel, and has come to represent many of the issues involved in the larger Israeli-Arab conflict. The public discourse in Syria and Lebanon on this subject has been driven by passion and politics rather than history and facts. Shebaa Farms serves as an example of how a place can become politically charged and important for no reason other than the meanings that people there assign to it.

## Geography and Attempts to Map the Area

The Shebaa Farms area is on the southwest ridge of the Herman mountain range, 14 kilometers in length, and consists of 14 agricultural properties.[1] It is near the border between the Golan Heights in Syria, and the Lebanese border,

***Samantha Borzi-Hedges*** *graduated in May 2013 with an M.A. in Statecraft and National Security Affairs with a focus in Public Diplomacy and Political Warfare. She received her BA in International Affairs with a minor in History from James Madison University in 2006. Mrs. Borzi-Hedges currently works within the office of the Deputy Assistant Secretary of Defense for Space and Intelligence.*

and does not hold a special place in history or geographic uniqueness.[2] The water in this area has been mentioned, however unconvincingly, as one of the reasons for contentions over the area.[3]

Several attempts were made prior to the 1980s when the conflict became prominent, to clearly demarcate Shebaa Farms and decide once and for all who the area belonged to. Though attempts at resolution were made, they were not considered serious, because it just was not an important enough issue to either Syria or Lebanon. Often, the various maps of the time ignored the de facto, political reality on the ground. The French occupied Syria from roughly the end of WWII until 1944. French maps located the Syrian border along the watershed in the area of Mount Herman, making Shebaa Farms part of Syria.[4] But the French never officially surveyed the area or clearly demarcated the area. After the French left, the issue was largely ignored until the 1950s when Syria took control of Shebaa Farms.[5] The area continued to officially belong to Syria until the 1967 war with Israel, in which Israel seized the Golan Heights and parts of southern Lebanon, including Shebaa Farms.[6] In 1974, when the Syrian-Israeli Disengagement Agreement was signed and UN forces were deployed to the Golan Heights, the existing maps were accepted by Lebanon, Syria, and Israel, placing Shebaa Farms once again in Syria.[7] During the course of this official undertaking to clearly demarcate the territory, the people of Shebaa Farms paid taxes to the Lebanese government, held Lebanese citizenship, went to Beirut to conduct official business, held deeds to the land going back generations, and generally thought of themselves as Lebanese.[8] Essentially, the maps and government officials in Israel, Syria, Lebanon and the UN all ignored the people in this area. The official maps did not match reality. This is always a dangerous situation because the people who identify themselves as Lebanese did not care what the maps said—they did not consider themselves Syrian.

The United Nations in practice has also agreed that this territory belongs to Syria. UN Security Council Resolution (UNSCR) 242 declares Shebaa Farms part of Syria.[9] The official maps that Lebanon has routinely approved with the renewal of UN Interim Force in Lebanon (UNIFIL) mandate have continued to show Shebaa Farms as part of Syria.[10] Then-UN Secretary General Kofi Annan acknowledged the ambiguity of the area and suggested using the blue line which separates UNIFIL and UN Interim Force in the Golan Heights (UNDOF) placing Shebaa Farms in the Golan Heights as part of Syria.[11] The issue remained relatively quiet until 2000 when the UN declared Israeli withdrawal from Lebanon complete.[12] Israel, however, remained in Shebaa Farms. Therefore, by the UN declaring Lebanon free of Israeli forces, it once again placed Shebaa Farms in the Golan Heights territory making the territory de facto Syria's. After 50 years of a hiatus on the issue, three weeks after the official withdrawal of Lebanese forces, the Lebanese made an official claim to the UN that the Shebaa Farms territory was in fact Lebanese, and the UN Secretary General made a note of Lebanon's new position.[13] Since then, Syria has made public statements that the area belongs to Lebanon, but refuses to make this position official.[14]

Everything official from the UN to Lebanese approving official maps places Shebaa Farms firmly in Syria, but everything on the ground makes the territory Lebanese. The issue gained prominence in 2000 when Hezbollah, the terror group dedicated to the destruction of Israel, brought the issue to the forefront of the conflict with Israel by staging attacks in Shebaa Farms on Israel and pledging to regain *all* Lebanese territory, including Shebaa Farms. Since then, the conflict has escalated and become one more obstacle that lies in the way of achieving peace between Israel and its Arab neighbors.

## Shebaa Farms Makes the Headlines

The general ambivalence that had surrounded the issue quickly evaporated in October 2000 when Hezbollah captured three Israeli Defense Force (IDF) Soldiers from the Shebaa Farms area. Hezbollah's stated goal in this operation was to exchange the Israeli hostages for Lebanese hostages, which eventually took place, in a grossly uneven deal favoring Hezbollah. In April 2001, another IDF soldier was killed in the area and Israel responded by firing tank and artillery shells into suspected Hezbollah hideouts. Israel also used fighter jets to attack targets in southern Lebanon. When, in August 2003, Hezbollah fired into Israel with rockets, anti-tank missiles, and mortar shells; Israel also responded to this attack.[15] This volley of attacks has continued several times since 2000. A region that was long forgotten suddenly became one of the main places where Hezbollah was to make its stand.

## Hezbollah's Point of View

One of Hezbollah's main objectives is to ensure all IDF soldiers leave Lebanon. From the Israeli and UN point of view, this happened in 2000 when the Israelis withdrew all the way to the Golan Heights. Hezbollah, on the other hand, believes that Shebaa Farms is Lebanese, and not part of the Golan Heights, and therefore it has a legitimate reason to continue attacking Israel. Created in the 1980s, Hezbollah did not take not of Shebaa Farms until Israel was declared to have completely left Lebanon.[16] This begs the question whether Shebaa Farms has true value, or whether it is a mere political tool of Hezbollah and a source of legitimacy for its continued existence.

Hezbollah leader Hassan Nasrallah has placed the issue of Shebaa Farms at the forefront of his rhetoric. He believes it shows a clear violation of Lebanese sovereignty. Nasrallah claims that "Hezbollah is a resistance movement not a militia," and that the identity of Shebaa Farms was Lebanese, therefore Hezbollah's resistance is legitimate. In an interview with New TV, Nasrallah makes known his disagreement with UNSCR 1701, which called for a ceasefire between Hezbollah and Israel, because he believes Hezbollah has a right to fight Israel *wherever* it occupies Lebanon.[17] Nasrallah has made several other statements similar in nature. He does not mention in any of them any sort of strategic value to Shebaa Farms or any reason for fighting in the territory other than Hezbollah's right to do so as a resistance organization. Hezbollah claims it is a resistance movement, which means it needs something to resist. Israel is gone from Lebanon, but since Hezbollah's creation until today, Hezbollah has gained steady power and popularity within Lebanese politics, which is something it does not want to lose. Hezbollah leaders also do not want to give up their arms; this means they need something to resist and Shebaa Farms offers the perfect excuse.

After the Hezbollah attacks in 2000, Kofi Annan declared that Hezbollah had violated UNSCR 425, the resolution that Israel had fulfilled when it left Lebanon in 2000. Lebanese President Emile Lahoud rejected this criticism, giving Hezbollah de facto approval for its actions and also demanded that Israel withdrawal from Shebaa Farms.[18] Suddenly, an issue that had not really mattered for years was being hotly discussed not only by a radical militant group, but also by a Lebanese government giving legitimacy to Hezbollah.

The leader of Lebanon's Druze community, Walid Jumblatt, feels that Shebaa Farms is an issue invented by Syria and Iran to give Hezbollah an excuse to continue fighting Israel and maintain influence.[19] There is some evidence for this statement. Even though the territory was taken in 1967, few Lebanese had heard of the territory until 2000, and they do not have any clear conception of where Shebaa Farms is or what it looks like.[20] There are a number of reasons why the average Lebanese citizen would be willing to support a cause that keeps Hezbollah alive. Most citizens feel that Hezbollah guarantees their safety and that the group is the reason Israel withdrew from Lebanon.[21] The Lebanese confessional system of government is a complete disaster and inept at dealing with Israel and Syria. Hezbollah has gained popularity, among Shia especially, as a party that battles Israel and brings some semblance of strength to Lebanon. Hezbollah is viewed as the one group capable of fighting the Israelis and maintaining Arab dignity in the face of perceived Zionist pressure. Therefore, citizens are willing to care about the issue of Shebaa Farms in a way they never did before, if it gives Hezbollah staying power and pride in the fact that a Lebanese group is able to take on the much more powerful Israel. Though as violence has escalated, popular support for Hezbollah has begun to diminish;[22] nonetheless, Hezbollah still enjoys enough support to be a main part of Lebanese politics.

The issue of Shebaa Farms also has a larger theme which can be seen in many aspects of Israeli-Palestinian conflict. For example, most Arabs view this issue as further proof of Israel's expansionist nature.[23] Many believe, incorrectly, that Shebaa Farms is being settled by Ethiopian Jews. Lebanon also claims that Israel is making substantial profit from wine production in this area, while in fact; there are zero vineyards in the territory. Shebaa Farms is also a local pilgrimage site because it is supposedly the location where the Covenant of Pieces was made between God and Abraham; therefore, this land has to be returned to Muslims.[24]

These issues draw a strong parallel to the larger Israeli-Palestinian conflict. The Jewish settlements in the West Bank are one of the major stumbling blocks in the peace process, so the idea that Israelis are once again settling Jews in Muslim territory is a primary offense. Israelis occupying what is viewed as a pilgrimage site is similar to the lack of access Muslims have to their holy sites in Jerusalem. In addition to the parallels listed above, Israel, in fighting Hezbollah, is also fighting Iran and Syria. Syria is Hezbollah's facilitator while Iran is its ideological and financial benefactor.[25] Even in this small battle, Israel is taking on more than one Arab nation and seeing, again, just how much these nations want Israel destroyed. How can Israel ever hope to reach a solution with Lebanon while Syria and Iran continue to meddle in affairs? Overall, the Arabs simply do not trust the Israelis and vice versa. This disagreement over such an insignificant territory has become symbolic of the larger fight for Palestinian independence and Israel's security problem.

The Lebanese truly believe that are fighting a legitimate fight to free the entirety of their territory from Israel, led by Hezbollah. They do not value this area for any particular strategic reason, nor do they live there at this time. Still, the area has become a symbol for everything wrong with Israel and its perceived continued oppression of Arabs. Because of the magnitude of Shebaa Farms as a symbol, it is likely the fight will continue—because when the Israelis leave Shebaa Farms, it will be a major political victory for Hezbollah.

## Israel's Point of View

The Israelis feel that they fulfilled their obligation to the UN and fully withdrew from Lebanon in 2000. They feel that Shebaa Farms belongs to Syria and is part of the Golan Heights.[26] The story of Shebaa Farms holds many parallels for the Israelis to the Arab-Israeli conflict just as it does to Lebanon. For Israel, the idea that Shebaa Farms belongs to Lebanon is a lie that serves to legitimize to Hezbollah.[27] It is further proof to Israel that the Arab world has no intention of ever accepting Israel as a state, due to the continued attacks and Hezbollah's zero-sum rhetoric.[28] The Arab League has also claimed the territory as belonging to Lebanon and blames Israel for the violence starting in 2000.[29] This is further proof to Israel that it simply will never have support from an Arab nation that Arab countries will always rally together against Israel.

In a July 2006 UN Security Council meeting, Israeli's ambassador to the UN Dan Gillerman said that the only reason Israel was in Lebanon was to protect itself from Hezbollah's blatant acts of war. He also charged Hezbollah mosques with spreading hatred toward Israel. Once again, Gillerman confirms the Israeli view that Israel has left every inch of Lebanon and that therefore these attacks are unjustified.[30] This incident confirms one of Israel's worst fears, that even if they give into Arab demands, as they did by leaving Lebanon, they will never be secure. Ultimately, the Israelis do not believe that Shebaa Farms belongs to Lebanon but to Syria; therefore if they want it back, they must come to some sort of agreement with Syria, not Israel.[31] This view has been largely backed by the UN and the US.

As is the case for Hezbollah, this is largely a political issue for Israel, not a piece of strategic ground. If the Israelis give into Hezbollah's demands, they are giving in to a terrorist organization that relies on violence. This sends a dangerous message to the rest of the Arab world that if Israel is bothered enough, it will give in. Israel also does not trust Hezbollah, or its backers Syria and Iran, enough to believe that if the Israel's leave Shebaa Farms the violence will end. Hezbollah is there to stay and its main goal is the destruction of Israel, whether it is occupying Shebaa Farms or not. Israel, thus, has nothing to gain by leaving Shebaa Farms.

## Conclusion

Shebaa Farms is not located in a strategic location, it does not hold a special place in history, nor is it an important economic center. In fact, there is nothing spectacular about it, and yet, it has managed to become one of the hottest political issues in the Arab-Israeli conflict. It serves as an example of how a place can become politically charged for no reason other than the meaning people assign to it.

The UN, U.S., and Israel all agree that Shebaa Farms belong to Syria. Syria and Lebanon agreed with this position until 2000. The dispute over Shebaa Farms had technically been ongoing for years before, but did not matter in the grand scheme of things until Hezbollah needed an excuse to continue fighting Israel, and a reason to keep its weapons. Shebaa Farms offered that excuse because, even though the dispute had been ignored for so long, the seeds for conflict were already present. Israel, for its part, will continue fighting for Shebaa Farms because resistance is Israel's automatic response to Arab threats. Israel is threatened from all sides and must live in a constant state of heightened vigilance. Israel can never allow itself to appear weak or appear to give into what it views as terrorist demands. Israel simply will not withdraw from Shebaa Farms merely because Lebanon claims the area as its own. Syria needs to officially proclaim that Shebaa Farms is, in fact, Syrian, before Israel will leave. Yet, Syria will never officially do so because such a declaration would take away legitimacy from Hezbollah, a group it backs. It's a classic Catch-22 where neither will budge or make the first move, and so the conflict continues without an end in sight.

Ultimately, Syria and Lebanon will have to prove one way or the other to whom the territory belongs. Then Israel will have to work out the details of withdrawal with the victor, while assuring its own security. If Lebanon is the victor, Israel will also have to negotiate with Hezbollah, something it has been unwilling to do thus far. It is likely that negotiations with Hezbollah will yield very little as long as Hezbollah is under the financial thumb of Iran, however, Hezbollah is a necessary part of the solution, as its leaders are the ones keeping the issue alive. Most likely, this issue will not be resolved until the Palestinian issue is resolved because that peace process will likely deal with the 1967 territories of which Golan Heights is a part. Once this is resolved, Shebaa Farms will likely go back to being an unimportant, inconsequential piece of land—but this time it will have a clearly demarcated border.

## Notes

1. Asher Kaufman, "Understanding the Shebaa Farms Dispute: Roots of the Anomaly and Prospects of Resolution," *Palestine - Israel Journal of Politics, Economics, and Culture* 11, no. 1 (April 2004): 37.

2. Reuven Erlich, "Raising the Issue of the Sheba'a Farms in the Proposed American-French Security Council Draft Resolution for Ending the Fighting: Background Information and Significance," *Intelligence and Terrorism Information Center at the Center for Special Studies*, August 9, 2006, 4.

3. Allah Hasan Nasr, Nicholas Noe, and Nicholas Blanford, *Voice of Hezbollah: the Statements of Sayed Hassan Nasrallah* (London: Verso, 2007), 240.

4. Asher Kaufman, "Who Owns Shebaa Farms? Chronicle of a Territorial Dispute, "*The Middle East Journal* 56, no. 4 (Autumn 2002): 594.

5. Asher Kaufman, "Understanding the Shebaa Farms Dispute: Roots of the Anomaly and Prospects of Resolution," 38

6. Reuven, 4

7. Asher Kaufman, "Understanding the Shebaa Farms Dispute: Roots of the Anomaly and Prospects of Resolution," 38

8. Habib Battah, "Letter from Shebaa," *The Washington Report on Middle East Affairs* 27, no. 9 (December 2008): 43

9. Robert G. Rabil, *Embattled Neighbors: Syria, Israel, and Lebanon* (Boulder, Co.: Lynne Rienner Publishers, 2003), 271

10. Augustus R. Norton, "Lebanon's Malaise," *Survival* 42, no. 4 (Winter 2000-2001): 50.

11. Asher Kaufman, "Understanding the Shebaa Farms Dispute: Roots of the Anomaly and Prospects of Resolution," 37

12. Battah, 42

13. Kaufman, "Who Owns Shebaa Farms? Chronicle of a Territorial Dispute," 592

14. Battah, 43

15. Anthony H. Cordesman, "Lebanese Security and Hezbollah," *Center for Strategic and International Studies*, July 2006, 23-25

16. Augustus Richard Norton, "The Role of Hezbollah in Lebanese Domestic Politics," *The International Spectator* 42, no. 4 (December 2007): 480

17. Nasr, Noe, and Blanford, 245, 367, 383

18. Norton, 42

19. Nasr, Noe, and Blanford, 387

20. Battah, 42

21. Cordesman, 17

22. Norton, 480.

23. Kaufman, "Understanding The Shebaa Farms Dispute: Roots of the Anomaly and Prospects of Resolution," 37

24. Kaufman, "Who Owns Shebaa Farms? Chronicle of a Territorial Dispute," 594-595

25. Edward P. Djerejian, "From Conflict Management to Conflict Resolution," *Foreign Affairs*, December/January 2006, 3.

26. Battah, 43

27. Kaufman, "Understanding The Shebaa Farms Dispute: Roots of the Anomaly and Prospects of Resolution," 37

28. Kaufman, "Who Owns Shebaa Farms? Chronicle of a Territorial Dispute," 594

29. Rabil, 276

30. United States., Congress., Senate., *Assessing the Strength of Hezbollah: Hearing before the Subcommittee on Near Eastern and South and Central Asian Affairs of the Committee on Foreign Relations, United States Senate, One Hundred Eleventh Congress, Second Session, June 8, 2010.* (Washington: U.S. G.P.O., 2010)

31. Ibid

# The Dragon and the Tiger: Countering Chinese Policy of Uniting the Smalls

Darren Fazzino and Max Macrides

The first weapons were human fists. Stones augmented these; then pointed stones; then stones attached to sticks. These became spears; which became compound bows; which became crossbows that yielded to firearms; and these have been developed since. Kinetic action is now so efficient that direct state-on-state conflict is an unwieldy cudgel of last resort. The perpetual avoidance of state-on-state conflict by most of the world creates a buffer zone in which a state can have aggressive action without much fear of triggering a conventional war. China operates in this zone because it is a regional power with an agenda that does not match the status quo, and no one can challenge it. Mitigating China's activity in this sphere needs a balance-of-power approach to create an offshore regional counterbalance and deny it strategic allies.

China's rhetoric of a peaceful rise is not congruent with its actions. Its economic attack on the Philippines and Japan via the Spratly Islands and Senkaku Islands respectively, its ongoing militarization, and its policy of aggressive industrial

***Darren Fazzino*** *is a double M.A. candidate at The Institute of World Politics in Statecraft and National Security Affairs (specializing in Defense Studies) and in Strategic Intelligence Studies. His interests include asymmetrical naval warfare, political influence operations and foreign intelligence services. He expects to graduate in December 2013.* ***Max Macrides*** *graduated with an M.A. in Statecraft and National Security specializing in Public Diplomacy and Political Warfare from The Institute of World Politics in May 2013. He has interests in unconventional and asymmetrical strategy, and plans to work within his specialization and interests.*

espionage all constitute the actions of a nation that is concerned with uprooting the current international order instead of finding a place in it. The South and East China seas are devoid of powerful counterweights to resist this geopolitical aligning. The only force of note is Japan, which is still under Article 9 of the Japanese constitution, which prohibits deploying forces outside Japan despite talks of repealing the article.[1]

A natural counterbalance to China would be India. While there is not a long and heartfelt relationship between India and the U.S., India is a good strategic partner for countering China's rise. Forcing China to divide its attention across two oceans means that the straits that connect the Indian and Pacific oceans obtain massive importance and reduces the attention China can pay to the Western Pacific. China's biggest interest is to secure the materials that fuel its growing economy. India also sits astride the major shipping lanes that feed the Chinese economy. China needs India to be distracted, while the U.S. needs India to have strategic freedom to challenge these shipping lanes.

India currently lacks the strategic flexibility it needs to be an effective counterweight to China. Part of China's foreign policy is seeking to restrict India's strength by befriending nations on its border. Strategically negating India by creating northern distractions gives China greater access to the Indian Ocean and its African supply roads. Leaving India as an active player means that China must have other routes by which to ship its resources around the Indian subcontinent. Each needs the other to be distracted, or suffer strategic inflexibility.

This policy of geopolitically upsetting India can serve to alter India's operational calculus. The states involved in this are Pakistan, Nepal, Bhutan, Bangladesh, and Sri Lanka.[2] All of these states have warming relations with China and poor to decent relations with India. All have one thing in common: They share a border with India and together serve to draw a ring around two thirds of India's northern territories.

Chinese-Pakistan relations have been historically good and continue to be so. China is one of Pakistan's biggest backers, so long as Pakistani insurgents do not plague western Chinese provinces.[3] Gwadar, one of Pakistan's largest ports, is now Chinese-run, and an ever increasing infrastructure network ties this port to Afghanistan, northern Pakistan and Western China.[4] Eighty percent of Pakistan's population sees China as a friend, whereas only 9 percent hold the same opinion about the U.S.[5] India must keep this axis in mind when making decisions. While individually these states are not a direct problem, together they pose a serious headache for anyone in New Delhi contemplating a new foreign policy strategy.

China's relations with Nepal are based on tying Nepal into China's economic circuit. Anti-Indian opinion is present in many editorials in the major papers.[6]

There is a large voice for Nepal to realign itself with China, and away from India. Kathmandu has elements of a Maoist government who have recently embraced China's model of socialism with capitalism. The nation has the highest paid and least productive workers and sees China as a way out of these problems. It is a nation seeking to expand and reform itself and the opportunity for India to make something of this opportunity is huge.

Bhutan is harder to draw into a Chinese orbit due to a tense relationship. The 1962 Sino-Indo war strained relations between China and Bhutan and they have not yet recovered. Bhutan identifies with Tibet, which is currently under Chinese control. India has a practical monopoly on all Bhutanese trade, accounting for over 90 percent of all trade from Thimphu.[7] China is trying to increase goodwill in Bhutan by creating economic ties and alternative markets other than India, as is evident by the increase in road construction linking China and Bhutan.[8]

Nepal and Bhutan represent geopolitical chess pieces far more than they do valuable trading partners. These two regions border the Indian province of Sikh, which is one of the few direct gateways between China and India, and is India's link with its eastern provinces. Bhutan also provides a way around the Tawang chokepoint, located in Upper Pradesh and home to some of the heaviest 1962 fighting. The economic opportunities for these countries are limited but their geopolitical meanings are enormous. India and China cannot be secure in their borders if these states fall into another's orbit.

China and Bangladesh have enjoyed very close ties since 1975. There is close economic, political and military cooperation between the two countries. India shares a close relationship with Bangladesh as well. China is aggressively vying for influence here by economic rail ties to Myanmar and by opening a new port.[9] This presents a problem for India, as Chinese warships will have a friendly port in the Bay of Bengal, which has historically been Indian dominated waters and threaten Indian ports. China seeks to reinforce its supply lines and find an alternative to the Straits of Malacca. It is imperative for India to deny China this opportunity.

Sri Lanka and China are tightly bound. Chinese technical skill, military hardware, and economic development and aid continue to permeate Sri Lanka. In November 2012, a $750 million infrastructure investment project was signed by China to help Sri Lanka develop.[10] Located east of India's southern tip, Sri Lanka holds immense political value, as it dictates entrance to the Bay of Bengal, where China and India both have a naval presence and access to China's overland rail link for energy transport through Myanmar,[11] which circumvents the Malacca Strait.

Beijing and New Delhi are caught in an unfortunate security dilemma. Beijing believes the world is becoming more multipolar, and seeks to accelerate that trend.[12] A 2009 suggestion by China that America recognize India as its "zone of influence" raised considerable eyebrows in New Delhi, as this would be turning the region on its head.[13] More concerned with Washington, Beijing sees New Delhi as a puppet rather than an independent actor and believes it can neutralize this arm of American foreign policy.

Because of this competition, the United States faces a crisis. Either it must act to pry these nations from China's orbit or it must abandon India as a strategic ally, as a New Delhi entangled with its northern neighbors cannot counterbalance China. Traditional U.S. diplomatic efforts have fallen short in altering Chinese behavior.[14] Therefore, the U.S. must turn to different methods of diplomacy and international political action to adapt to this ongoing challenge.

China's strategy regarding these nations is overt political warfare to corral these states, or at least make them unfriendly toward the U.S., often by use of business contacts and economic ties. Public diplomacy and political warfare are two underused tools of the U.S. diplomatic function, yet constitute a very important part of international affairs. They are an acceptable part of the zone between kinetic action and apathy and thus are vital tools in the new international order. Ongoing American reluctance to use these tools is a massive hindrance to foreign policy objectives.

China's India strategy is contingent upon tying India down in a security morass with its northern neighbors. Denying Chinese influence in these countries will undo the security of China's western borders and, in turn, will pull attention from the Pacific at a time when tensions are very high with Manila, Tokyo, Hanoi, and other states. China requires India to be wrapped up in its northern security affairs, distracting it from growing Chinese power in the Indian Ocean. China's trade lines run directly across the Indian Ocean making them vulnerable to interdiction by the Indian navy.

The strategy to deny China those areas resides in giving these populations an alternative narrative on Chinese actions and giving regional governments an alternative to a Beijing-centric order. Any discontent sown between these capitals and Beijing allows India strategic maneuverability. More regional flexibility is good, as it draws more of China's attention to the Indian Ocean and away from the western Pacific.

India is the strategic cornerstone to countering a rising China. The first step will be forming closer diplomatic ties with India, since a strategically free India that does not consider American wishes could be very dangerous. This will ensure that there is coordination of strategy and unity of purpose. It will also provide a clear indicator to China that the United States is cognizant of its strategy and seeks to free India from such entrapment. Closer ties also suggest to other nations in the region that the U.S. is serious about its commitments and the "Asian Pivot" in general.

Public diplomacy will serve as half of a two-pronged method to strip China politically of these territories, with the other prong being political warfare. While not all democratic, these countries' public moods are alterable and do impact government opinion and action. They do have open presses, which can be used as launching pads for ideas and conversations. Chinese ambassadors print pro-China editorials and letters in all of these major papers and there is no pro-West, pro-India, or pro-America narrative to challenge the viewpoint.[15] One major regional news source is China's news service, which occasionally propagates an anti-American Chinese jingoism.[16] Thus, Beijing tells the American story. The lack of an alternative narrative distorts the information environment. Populations form misconceptions based on this compromised system.

Two actions will shed some light to this issue. First, providing short-wave radio broadcasts of unaltered news, history, and culture will ensure that these populations have accurate frames of reference when dealing with Chinese propaganda. Informational leaflets demonstrating Chinese strategy through uncaptioned pictorials will also provide a direct communication to educate the general populations of these countries. These claims will have to be dispelled by China, and the result will be some awkward questions or the uncovering of uncomfortable truths. Diasporas should be contacted to provide content and create a strong link with the locals.

Throughout, America needs to be working with both the governments and the people in order to establish conditions that can resist Chinese expansion. Public diplomacy must be factual about offering to fix problems through appropriate local channels, and Chinese abuses in Africa, religious massacres, and general imperialism should be made known as well. The message is quite simple: Be our friend, or China's slave.

China supplies much of the money being used to rebuild Sri Lanka, without the strings that Western money can have. This may limit Western financial aid, but advisors seen handing out the aid could go a long way toward increasing friendly influence in the reason while undermining Chinese Strategy.[17] Similarly, America can offer medical diplomacy incentives as well as logistical support. Therefore an international coalition anchored and managed by the U.S. and India should offer joint U.S. military medical diplomacy and provide support for Sri Lanka's restructuring and refugee resettlement, as well as India-centric infrastructure projects. Once home to a booming tourist industry, the Tamil Tigers and the 2008 credit crunch constricted growth in the country. Helping restore the tourist industry would create much good will in the Sri Lankan business community.

Bhutan is a prime candidate for medical diplomacy and an Indian-centric infrastructure. Bhutan needs an official embassy and diplomatic relations to create an outlet for legitimacy and close discussion. America can work through India to provide improved high-altitude technological construction aid, especially hydropower. The country requires a lot of work to get adequate infrastructure. Providing infrastructure aid will ensure that Bhutan can grow into a stronger trade partner and could provide a powerful geopolitical buffer to bog down a Chinese military excursion. In addition, Bhutan's total external debt is $1.5 billion.[18] While this represents 55 percent of Bhutan's GDP, India and the U.S. could wipe that out and provide a boost to the Bhutanese economy.[19] The dangers of debt in an economy are well known.

In Bangladesh, opportunities include widespread anti-flooding projects funded by India and the U.S. Helping with infrastructure, medicinal diplomacy, and refugee issues will go a long way toward building the country. Advisors can help mitigate the chronic mismanagement present in Bangladesh. Bangladesh's Muslim majority opens it to propaganda involving the Uigher massacres by the Chinese.[20] These segments can be used as a tireless, irate minority to drive parts of Bangladeshi domestic policy away from China.

After securing these states, attention can turn to the harder target: Nepal. This country has a virulent anti-Indian tilt, but suffers severe economic problems brought about by a Communist revolution.[21] If the other countries are doing well, they could serve as models for Nepal, which may adopt a neutral stance. Economic growth and medical diplomacy are tools that cannot be refuted and are hard to label as imperialism. Nepal may not fall into the Indian orbit, but neutrality is enough to upset Chinese plans. Tying Nepal into an India-centric economic model will cause Nepal to be less accommodating to Beijing.

The unsettled state remains Pakistan, containing a rabid anti-Indian sentiment. There is nothing that can be done to fix this reality. India must seek strategic depth with Pakistan by involving itself in Afghanistan. This will divert the Pakistanis attention from their eastern border, which is the best outcome given the states involved. Trying to drag Islamabad into New Delhi's orbit is folly; all that can be done is a neutralization of its effects.

The overarching objective for this strategy is to provide an alternative narrative to the one disseminated by Chinese state TV, and to back it up with material goods and services as demonstrations of irrefutable American good will. All nations need roads and infrastructure to function, and if these can be built with American money and Indian advisors in these nations, all the better. The goal is to increase the economies of these nations and orient them to an India-centric economic model to deny China inroads and infiltration opportunities by businessmen who double as agents of influence. Infrastructure projects will help accomplish this because it will be easier for companies to get to India than to China, and they should be more inclined to do business there.

This new narrative will give the populations another means by which to hear of the outside world. While the new narrative may not drive the countries into a pro-American fervor, it will serve as a counter to Chinese claims and propaganda and start debate in the countries. A neutral alignment is better than a pro-China tilt.

A plan like this is important for a few reasons: China's geopolitical western security is at risk, which is part of why it is pursuing these regions. These areas act like buffers between China and India. When aligned which one side or another, the Himalayan states can provide launching pads or just create headaches for either party. Controlling mountain passes is paramount to regional power. It is hard to project influence over an area if one cannot get there.

The more China is distracted from the Senkakus or Spratleys, the more stable the region is. Chinese rhetoric as of late is similar to Japanese rhetoric of the 1920s. The same jingoistic lines ring loud and clear in a region that remembers its history well. Forcing attention to other regions could have balancing effects in the U.S. as well, from which everyone would benefit. Halting realignment of the states that border India would be a helpful step in this direction. Warfare, once evolved from fists, stones, and sticks now can move to the acme of skill: information, diplomacy and strategic positioning—or regress to bombs, bullets, and tears.

(Endnotes)

1 Crowell, Todd, "Abe, Third Party Stir Japanese Politics," Real Clear Politics, http://www.realclearworld.com/articles/2012/09/17/return_of_abe_100239-2.html, (accessed February 3, 2013).

2 J. MOHAN Malik, *Pacifica Review*, *South Asia in China's Foreign Relations*, Vol. 13, No. 1 (2001): 73–90, http://sga.myweb.uga.edu/readings/sa_in_chinas_fr.pdf.

3 Mizrokhi, Elena, "China-Pakistan Relations: Not As Advertised–Analysis," *Eurasia Review*, July 31, 2012, http://www.eurasiareview.com/31072012-china-pakistan-relations-not-as-advertised-analysis/ (accessed January 31, 2013).

4 Shengran , Zhao, "Pakistan to transfer Gwadar Port's Management to China," *China Daily*, January 2, 2013, http://usa.chinadaily.com.cn/world/2013-02/01/content_16192162.htm (accessed January 31, 2013), Pant, Harsh V. Japan Times, "China's Gwadar plunge could constrain India." Last modified Feburary 23, 2013. http://www.japantimes.co.jp/opinion/2013/02/25/commentary/chinas-gwadar-plunge-could-constrain-india/

5 Pant, Harsh V, "The Pakistan Thorn in China-India-U.S. Relations,"The Washington Quarterly (Winter 2012), http://csis.org/files/publication/twq12winterpant.pdf (accessed Janu-

ary 31, 2013).

6 Regmi, Madan. TelegraphNepal.com, "Nepal: Pivotal for preservation of China's security interests." Last modified 2012. Accessed January 31, 2013. http://www.telegraphnepal.com/views/2012-09-22/nepal:-pivotal-for-preservation-of-chinas-security-interests.;SHRIVASTAV, Ariv. The Himalayan Time, "Anti-India and self-critical Nepali syndrome." Last modified 2013. Accessed January 31, 2013. http://www.thehimalayantimes.com/featured/year2067/inner_blogs.php?newsid=145.

7 Galey, Karma. The Centre for Bhutan Studies, "AN ANALYSIS OF PEGGED EXCHANGE RATE BETWEEN BHUTAN AND INDIA ." Last modified 2007. http://himalaya.socanth.cam.ac.uk/collections/journals/jbs/pdf/JBS_08_07.pdf.

8 MATHOU, THIERRY. Bhutan Studies, "BHUTAN-CHINA RELATIONS: TOWARDS A NEW STEP IN HIMALAYAN POLITICS." Accessed January 31, 2013. http://www.bhutanstudies.org.bt/pubFiles/19-Spdr&Pglt.pdf.

9 "The Begums and the Two Giants,"The Economist, January 9, 2013, http://www.economist.com/blogs/banyan/2013/01/bangladesh-and-its-near-abroad?zid=306&ah=1b164dbd43b0cb27ba0d4c3b12a5e227 (accessed February 1, 2013).

10 The Indian Express, "China, Sri Lanka to boost defense relations." Last modified 2012. Accessed February 1, 2013. http://www.indianexpress.com/news/china-sri-lanka-to-boost-defence-relations/1030109.

11 Perlez, Jane. "Laos Could Bear Cost of Chinese Railroad." *New York Times*, New York editionJanuary 2, 2013. http://www.nytimes.com/2013/01/02/world/asia/china-builds-a-railroad-and-laos-bears-the-cost.html?pagewanted=all&_r=0 (accessed February 3, 2013).

12 Jamestown Foundation, "China's 2010 National Defense White Paper: An Assessment." Last modified 2012. http://www.jamestown.org/single/?no_cache=1&tx_ttnews[tt_news]=37839.

13 Pant, Harsh V. CSIS, "The Pakistan Thorn in China—India—U.S. Relations."

14 Loy, Irwin. Voice of America, "South China Sea Dispute Flares at ASEAN Summit." Last modified 2012. http://www.voanews.com/content/south-china-sea-dispute-flares-at-asean-summit/1548880.html.

15 Larson, Christina, "China's Fox News," *Foreign Policy*, October, 31, 2011, http://www.foreignpolicy.com/articles/2011/10/31/global_times_china_fox_news (accessed April 8, 2013).

16 Kazianis, Harry,"Beware of Chinese Jingoism," May 15, 2012, http://thediplomat.com/china-power/beware-of-chinese-jingoism/ (accessed April 8, 2013).

17 Wheeler, Thomas. The Diplomat, "China's Development Diplomacy." Last modified 2012. Accessed February 25, 2013. http://thediplomat.com/2012/03/04/china'-development-diplomacy/?all=true.

18 CIA, "CIA world fact book: Bhutan." Last modified 2012. http://ciaworldfactbook.us/asia/bhutan.

19 Federal Reserve Bank of St. Louis, "Central government debt, total (% of GDP) for

Bhutan (DEBTTLBTA188A)." Last modified 2012. Accessed February 3, 2013. http://research.stlouisfed.org/fred2/series/DEBTTLBTA188A.

20 BBC, "Turkey attacks China 'genocide'." Last modified 2009. http://news.bbc.co.uk/2/hi/8145451.stm.

21 Kreuttner, Timothy R. Small Wars Journal, "The Maoist Insurgency in Nepal and US Counterinsurgency Doctrine." Last modified 2009. http://smallwarsjournal.com/jrnl/art/the-maoist-insurgency-in-nepal-and-us-counterinsurgency-doctrine.

# How Iran's Geography Shapes Iranian Naval Strategy

Joseph Humphrey

Iran is situated at the crossroads of Asia and Europe. It is bordered in the south by the Persian Gulf and the Gulf of Oman, which are connected by the Strait of Hormuz. To the east, lie Pakistan, Afghanistan, and Turkmenistan, while it is bordered on the west by the Persian Gulf and Iraq. Iran's northern border is Armenia, Azerbaijan, Turkmenistan, and the Caspian Sea.[1] Iran's 70 million people make it the third-largest country in the Middle East (behind Egypt and Turkey) and the 18th-largest in the world.[2] In comparison to the United States, "if Iran were superimposed over a map of North America, it would extend north to south from Reno, Nevada, to Monterey, Mexico, and west to east from San Diego, California, to Amarillo, Texas."[3]

More noteworthy, however, is that Iran's oil reserves are the world's second largest and one of the world's major producers of natural gas.[4] Also noteworthy is that Iran's climate and topography are extreme in that they vary drastically, depending on the particular region of the country. As a result of being located near the Arabian, Indian, and Eurasian plates' convergent point and exactly where the Arabian and Eurasian plates meet, Iran is situated on what is known as the Iranian plateau. These plates are responsible for the Zagros Mountains along Iran's western coast and make Iran one of the most mountainous regions of the world.

The pressure between these tectonic plates also makes Iran a major earthquake

***Joseph Humphrey*** *earned a B.A. in Political Science in 2011 from Benedictine College in Atchison, KS before enrolling at IWP. He graduated with an M.A. in Statecraft and National Security Affairs with an emphasis in Intelligence in May 2013.*

zone.[5] The northern region of Iran is covered in thick rainforest. However, the high mountain ranges prevent rain clouds from reaching eastern Iran and it consequently consists of desert basins such as Dasht-e Kavir. The one major interior body of water that Iran posses is Lake Urmia, but it is saltier than Utah's Great Salt Lake and therefore, nearly useless.[6] Despite the rainforests and deserts, areas in the Zagros basin to the west experience severe winters with heavy snowfall and subzero temperatures.

All of this exemplifies the fact that Iran's geography is extreme, and when studying Iran, particularly its military and even more particularly its navy, it is important to be aware of this geography. This unique geography of Iran has shaped its naval capabilities and provided many interesting opportunities and challenges.

## History of the Iranian Navy

The Iranian navy, like the country itself, is shaped by its history and geography. Unlike many countries, "Iran does not have a long naval history."[7] Its existence only dates back to pre-Islamic revolution and to the discovery of Iran's petroleum deposits in the early 1900s. The discovery of these deposits and the subsequent potential for great wealth created the need for the Iranians to protect their maritime commerce. However, even with this new commerce, maritime protection was largely provided by the British as the Shah's navy operated under its shadow. This was the case until the British "stewardship" of the Persian Gulf ended in the 1970s and Iran took on a larger role in protecting the Persian Gulf sea lines and its merchant ships.[8]

Finally out from under the shadow of the British forces and with a surplus of oil revenue, the Shah expanded defense budgets and pursued plans to dominate the region's waters with the Shah's imperial navy, the Islamic Republic of Iran Navy (IRIN). Through relationships with the West, Iran purchased destroyers, frigates, and patrol crafts to fulfill its expanding mission. However, with the Islamic Revolution in 1979, the Shah was disposed, the ties to the West cut, and the Iranian naval aspirations were left unfilled. Although IRIN still existed, the new Islamic Republic under the leadership of Supreme Leader Ayatollah Khomeini developed the new Islamic Revolutionary Guard Corps Navy (IRGCN) and gave it the mandate to protect the revolution.[9] After severely weakening the IRIN by executing many of the senior commanders in attempts to eliminate Shah loyalists, IRGCN gained prominence in the Iran-Iraq War in the 1980s. Since that time, IRGCN has been favored by the leadership, who still view IRIN as potential Shah loyalists and have directed most of the funding to IRGCN.

As previously stated, prior to the 1970s, Iran depended heavily on Western support to develop its military capabilities and existed essentially under the shadow of the British stewardship. Because of this, it now lacks the defense industrial industry to produce its own arms and is therefore largely dependent on foreign arms suppliers such as North Korea, China, and Russia.[10] Iran is pursuing the means to be self-sufficient, but currently depends mostly on these foreign defense industries to produce and even maintain their forces. Iran similarly is confronted with the challenge of its mountainous geography in developing its naval capabilities. The Zagros Mountains, which run along the majority of Iran's rugged coastline, inhibit the ability to develop ports conducive to large vessels.[11] The Iranians have subsequently relied on smaller, faster vessels. According to the U.S. Office of Naval Intelligence, "while the IRIN is compromised primarily of aging Shah-era vessels," the IRGCN, the politically favored force, "has continued to purchase and construct new vessels over the years."[12] These vessels have been small and faster craft that are more suited for the IRGCN's mandate, a mandate that has subsequently developed into a "non-traditional" force that is focused on surviving threats while incorporating asymmetric defenses.[13] Iran's navel capabilities are not limited to small craft, though. In addition to smaller boats, Iran has developed a credible mining capability. However, "unconventional tactics" have been necessary in laying the mines because Iran has a limited number of conventional vessels capable of mine-laying.[14] The mines are deployed predominantly through the use of commercial and small vessels. Furthermore, Iran posses the Gulf's only submarine program and currently possesses three operational classes of submarines: the KILO, YONO, and NAHANG.[15] Iran purchased three KILO-class diesel-electric submarines from Russia in the 1990s and has developed seven YONO-class midget submarines with more in production. Iran also possesses one NAHANG-class midget submarine.[16]

The Iranian naval forces that exist today are a direct result of Iran's history and geography. Iran's petroleum reserves initiated the development of a naval force while the mountains and limited exactly what could be achieved. Similarly, Iran's historic revolution created the two distinct naval forces that it possesses today, while Iran's proximity to strategic geographic locations continues to shape its naval strategy. The connection is irrefutable.

## Strategy

Nearly every aspect of the current state of Iranian naval capabilities has been impacted by Iran's geography. The Zagros Mountains, the lack of inland waterways, and its close proximity to the Persian Gulf along with Iran's history have shaped both the IRIN and the IRGCN. Their strategy is similarly shaped by geography, particularly by the Strait of Hormuz. With the elimination of the Shah, his plans to dominate the regions waterways were ultimately terminated.[17] Generally, Iran's naval forces are used to protect "Iranian waters and natural resources, especially Iran's petroleum-related assets and industries."[18] However, the origin of Iran's naval strategy "can be found within the context of the Iran-Iraq War" before the distinction between IRIN and IRGCN was as obvious as it is today.[19] Both the IRIN and the IRGCN participated in operations during the war, but the IRCGN's small-boat tactics quickly set it apart and established IRGCN as a legitimate entity.[20] The success of the small boats further established the use of asymmetric tactics. In the tanker War (1984–1988), although conventional tactics were, in fact, employed, the ambushes and hit-and-run attacks by Iran's swift boats proved most successful. Due to these successes, not only did the IRGCN garner the majority of defense spending, but the IRGCN also decided that "Iran should plan to fight asymmetric war against enemies."[21]

Iran has tailored its naval defense strategies according to geographic location, with the most prominent being the Strait of Hormuz. Iran's naval leadership now believes that "today's threats across the world are sea-based and Iran needs to design naval forces and strategy to defend against them."[22] With Iran's close proximity to the Persian Gulf, it has the opportunity to make the Strait of Hormuz the key point of its strategy, and according to reports by the ONI, "Iran has achieved the capability of 'easily' closing off the Strait of Hormuz" if it so desired.[23] Control of the Strait gives Iran significant leverage not only in the Middle East, but also in the global community.

According to the U.S. Department of Energy, in 2008, the Gulf countries produced 29.8 percent of the world's oil supply and 29.1 percent of the natural gas. Most of this oil and natural gas was shipped through the Strait of Hormuz.[24] If the Strait of Hormuz were to be closed, these exports would have to be shipped over land. However, since the Middle East does not have the infrastructure to handle the quantity of shipments, supply would fall. The world, particularly industrialized nations, and the global economy would experience significant negative impact if the Strait were closed for an extended period of time. However, Iran would not be insulated from harm if the Strait were to be closed. Iran's oil exports are typically 75 percent of its total exports and near 30 percent of the country's GDP.[25] Closing the Strait would cause Iran significant economic harm and would not be undertaken lightly. Simply threatening to close the Strait or disrupt traffic is an effective strategy for Iran.[26] Iran understands, though, that the West would not long allow the closure of the Strait of Hormuz. With the entire industrialized world essentially dependent on the oil exports through the Gulf, the West would be quick to retaliate.[27] With this in mind, in addition to its campaign to become militarily self-sufficient, Iran has begun the process of increasing the range of its naval power so that it might address perceived threats without closing the Strait of Hormuz. Naval modernization, therefore, is one of Iran's highest defense priorities.[28].

IRIN, which has long suffered the problems of trying to maintain outdated Western-built vessels without Western support and adequate funding, has begun to "push operations further into the Gulf of Oman" and the Indian Ocean. These "extended patrols" will soon be supported by new naval bases that IRIN plans to develop by 2015 along the more geographically friendly coastline of the Gulf of Oman.[29] Iran's new emphasis on increasing capabilities is exemplified in its strategic military exercises and the procurement of strategic delivery systems for warheads.[30] While the IRIN increases its presences and influence outside the Persian Gulf, the IRGCN continues to employ asymmetric tactics with small, quick boats almost exclusively in the Persian Gulf. Although, the IRIN is beginning to develop and pursue mid-larger mid- to large-size vessels and retrofit the ones they already possess, IRGCN "has concentrated on acquiring and developing small, fast boats, some lightly armed and others armed with missiles and torpedoes, and will probably continue this trend."[31] Simply put, Iran's naval forces, although expanding, are unlikely to make drastic changes to their strategy.

**Conclusion**

Iran's naval forces have been and continue to be shaped by its geography and history. Iran will utilize its geopolitical position while continuing to explore further developments. Although Iran's strategy does not show signs of extreme fluctuation, there are two factors that will prove significant in the near future. With the toppling of Saddam Hussein, one of Iran's biggest deterrents of aggression in the region, Iran may prove to be bolder than in the past. On the other hand, the division in Iran's naval leadership structure may prove to be a source of weakness for Iran and its aspirations. Overall, Iran's geography and history provide both obstacles and challenges. If Iran is able to overcome the obstacle, it will be in a position to inflict significant influence over the region and the world.

## Notes

1 *Concise Atlas of the World*. New York: DK Publishing, 2008, 143.

2 Steven Ward, *Immortal: A Military History of Iran and its Armed Forces*, (Washington, D.C.: Georgetown University Press, 2009), 2.

3 Ibid, 5.

4 Ibid, 2.

5 Steven Ward, *Immortal: A Military History of Iran and its Armed Forces*, (Washington, D.C.: Georgetown University Press, 2009), 5.

6 Ibid, 5.

7 Iran's Naval Forces: From Guerilla Warfare to a Modern Naval Strategy." *Office of Naval Intelligence*,:1.

8 Iran's Naval Forces: From Guerilla Warfare to a Modern Naval Strategy." *Office of Naval Intelligence*,:2.

9 Ibid, 13.

10 Ibid, 13.

11 Steven Ward, *Immortal: A Military History of Iran and its Armed Forces*, (Washington, D.C.: Georgetown University Press, 2009), p. 2.

12 Iran's Naval Forces: From Guerilla Warfare to a Modern Naval Strategy." *Office of Naval Intelligence*,:13.

13 Ibid.

14 Ibid, 16.

15 Ibid, 18.

16 Ibid.

17 Iran's Naval Forces: From Guerilla Warfare to a Modern Naval Strategy." *Office of Naval Intelligence*,:2.

18 Ibid.

19 Ibid, 6.

20 Ibid, 1.

21 Iran's Naval Forces: From Guerilla Warfare to a Modern Naval Strategy." *Office of Naval Intelligence*,:7.

22 Ibid, 9.

23 Austin, Michael. "U.S. Navy v. Iran; The return of sea control?." *The Daily Standard*, December 17, 2009, 1.

24 Iran's Naval Forces: From Guerilla Warfare to a Modern Naval Strategy." *Office of Naval Intelligence*.:2.

25 Ibid.

26 Iran's Naval Forces: From Guerilla Warfare to a Modern Naval Strategy." *Office of Naval Intelligence*.:2.

27 Bodansky, Yossed. "Iran's Persian Gulf Strategy Emerges Through Its Recent Military Exercises." *Defense & Foreign Affairs' Strategic Policy*, January 31, 1994, 4.

28 Iran's Naval Forces: From Guerilla Warfare to a Modern Naval Strategy." *Office of Naval Intelligence*.:25.

29 Ibid.

30 Bodansky, Yossed. "Iran's Persian Gulf Strategy Emerges Through Its Recent Military

Exercises." *Defense & Foreign Affairs' Strategic Policy*, January 31, 1994, 1.

31 Iran's Naval Forces: From Guerilla Warfare to a Modern Naval Strategy." *Office of Naval Intelligence.*:25.

# Active Measures in Hollywood

## Richard Burns

Lenin's head of the Comintern, Grigori Zinoviev, declared that motion pictures "can and must become a mighty weapon of Communist propaganda and for enlightening of the widest working masses."[1] The Soviets understood the importance of film and, more important, popular icons and their influence on the masses: "[S]ome men and women whose every instinct rebels against the sound of a politicians' voice, are so conditioned that they are unable to resist when their favorite movie star whoops up an issue."[2] For these reasons, Hollywood became a target of Soviet infiltration for influencing the content of American movies as well for the advancement of other objectives. The active measures campaign against Hollywood was one of gradual development that relied on the gullibility and fears of people, and employed the tactics of subtleties and layers of lies and half-truths to conceal their involvement.

### The Players

Those involved can generally be divided between those actively controlled or influenced by the USSR, non-controlled communist sympathizers, and "useful idiots." Of those controlled by the USSR, the most important individual was

***Richard Burns*** *graduated in May 2013 with an M.A. Strategic Intelligence Studies M.A., concentrating in Counterintelligence. Mr. Burns received a B.A. in History with a minor in Arabic and Islamic Studies from the University of Kentucky. He is currently working as an intern at the National Consortium for the Study of Terrorism and Response to Terrorism collecting open source intelligence to construct organizational profiles of European crime groups and conducting research for the creation of a European architecture system to assess the threat of radiological material acquisition for use in terrorist acts.*

Willi Munzenberg. He was chosen by Lenin in 1921 to run all propaganda efforts against the west. It was through Munzenberg that the Soviets set up shop in Hollywood.[3] In fact, after the seventh Congress of the Comintern, Munzenberg's operations became the model for all U.S. and European activities.[4] Of his lieutenants, the most important was Otto Katz. Katz operated under the alias of Rudolph Berda and formed the front group Hollywood Anti-Nazi League. This group and many other organizations included the likes of Ella Winter, Donald Ogden Stewart, and Dorothy Parker. Through their social and professional networks they recruited others to their cause. Ernest Hemingway and Humphrey Bogart are examples of those wooed or tricked into speaking and engaging in activities that further advanced Soviet objectives.

The individuals mentioned are representative of the full array of people used. Donald Stewart has been described as easily manipulated, while Dorothy Parker was clearly a communist sympathizer, if not someone who received instruction from the Communist Party or Comintern.[5] Others, such as John Howard Lawson, would advance their agenda by serving as a jack-of-all-trades. He served as an attack dog to discredit and marginalize those who stepped outside party lines, organized protests of American policies, and inserted Communist Party propaganda and ideology into his work.[6]

## The Objective

Infiltration of the film industry for propaganda use was always an objective, but not always the main one. This active measures campaign could be characterized by a four-pronged approach. The first was to solicit funds from elites that were then channeled to the Communist Party USA (CPUSA) or Comintern coffers.[7] Another was to harmonize Hollywood's efforts with those of the popular front in Europe.[8] Here people like Boris Morros, an employee for Paramount Pictures, provided Soviet agents with employment and training for their film covers in the company's European offices.[9] These operations demonstrate the scope and breadth of activities and the people involved. A third goal was to get the average citizen to question the culture and political foundation of his own country. The Soviets, "never attempted to create an American Party or communist movement capable of even remotely challenging the constitutional power, as he [Stalin] would do in Germany . . . . The apparatus of American communism would be directed instead toward discrediting American politics and culture and assisting the growth of Soviet power elsewhere."[10] Lastly, the goal was to influence films through the subtle incorporation of ideology into the content of scripts and plays.

## The Methods

The best place to start is the actual climate and culture of Hollywood in the 1920s. Early on, Hollywood was not the glitz and glamour it is today. For those starting out, Hollywood was a cutthroat and lonely place to work with real class disparities between the "talent" and the rest of the workers. From the CPUSA's perspective, the "studio system was the perfect paradigm of monopoly capitalism."[11] The communists were successful at reaching young people by offering them a sense of community, a social life, and access to people who could advance their careers.[12] This sense of community and the working conditions played to the Soviets favor. Through human psychology, the Soviets advanced their agenda by highlighting and contrasting working conditions in a manner that validated Soviet ideology. The working environment was one that perfectly illustrated how communists described capitalism. This combination of factors opened the door to Hollywood. It may not have made someone a communist, but it certainly decreased the stigma attached to it.

The next step was to "seize on the most salient moral claims of the adversary culture" and promote the value of the grand experiment the Soviets were trying. [13] This step was partially based on the first already mentioned. An example of attacking the adversarial culture was highlighting the treatment of black Americans. This sad truth of the country's society allowed the Soviets and their sympathizers to take the moral high ground by declaring that communism championed true equality as well as addressing material inequalities.[14] All this was done covertly. Munzenberg's wife, Babette Gross, said, you never endorse Stalin, "you do not call on people to support the Soviets. Even under no circumstances. You claim to be an open-minded idealist. You don't understand politics, but you think the little guy is getting a lousy break. You are frightened by the racism, by the oppression of the workingman."[15] This was done with the hope that people would believe actions helping the Soviets were, "derived from the moral essential elements of human decency."[16] Or more succinctly, "communism was an intoxicated state of mind, a glow of inner virtue, and a sort of comradeship in super-charity, a way for the wealthy to posture as proletarian wage slaves."[17]

Furthermore, communists took advantage of the perceived-as-greater threat of fascism. Anti-fascism components helped achieve the goal of bilking money from elites as well as shift the focus away from the Soviets' own atrocities. This was accomplished through people and front groups like Otto Katz's Hollywood Anti-Nazi League. These goals are illustrated through an event in which Katz visited Hollywood under the alias of Rudolph Berra. During a fundraising event Katz portrayed an aristocrat who escaped Germany and was now telling the tale of how he fought the Gestapo. Katz correctly assumed that Hollywood could be manipulated with his story. With this speech, Katz set the ball rolling for the creation of the Anti-Nazi League in an attempt to allow those in Hollywood to "help" the oppressed in Spain and Germany.[18] This event demonstrates the importance of combating fascism in the minds of Americans. It further describes the naiveté of supporting communists in an effort to fight a "greater evil".

A German refugee, Salka Viertel, as well as others, knew who "Rudolph Berra" truly was, but kept silent.[19] The concept of combating a "greater evil" is validated by the words and actions of Viertel himself. It shows the willingness of some to overlook the flaws of communism and the USSR, or remain oblivious to possible Soviet connections to the League. In his book, Viertel describes an instance where a friend told him he was leaving the Anti-Nazi League because it was controlled by communists. Viertel had known Otto Katz as a communist in Berlin, but this knowledge and that of his friend's did not change his attitude. He did not believe it to be true because Katz couldn't be a socialist; he's fighting the fascists in Germany.[20]

However, none of this could have been possible without the help of "useful idiots." This category includes those ignorant of the true agenda of Hollywood front groups, those too good-natured to see the truth, and those who believed in communism unconditionally. They may not have been controlled or given explicit instructions by Soviet agents, but they stood on the side of communism or allowed it to march on. Two examples of this are Arthur Miller and his play *The Crucible*, and Donald Ogden Stewart. The "dupe" could be described as the third-generation product of this active measures campaign. He was also a tool in the continuation of it. At the first level are those who ran the CPUSA, a front group, or were part of the Comintern. Examples of these people are Willi Munzenberg, Otto Katz, and perhaps even John Lawson. They would tell others how best to pursue this agenda (fighting fascism or helping the little guy) without attracting attention. These people may not have known Katz and Munzenberg were controlled agents and simply followed their suggestions. These people could be classified as "dupes" but the next level is probably where the most success was. This can be seen in *The Crucible*.

In the 1950s, many believed the play was really about anti-communist "witch hunts" rather than the Salem Witch Trials.[21] Miller declared that it wasn't about anti-communist witch hunts even though the House Un-American Activities Committee documented his communist sympathies. As he declared his innocence, people came to his defense. These people could be classified as the truest of "dupes." They were tired of chasing communists or were simply defending their friend. In doing so, they prevented Soviet-implanted subtleties from being confronted. This sentiment is best described through the words of Lauren Bacall: "[T]hey had been foolishly naïve, headstrong, emotional, and that they had hastily strolled into something 'we knew nothing about.'"[22]

Donald Stewart was the perfect dupe. He supported communism but distanced himself from the USSR and Stalinism by tailoring his beliefs to suit his opinion. He described his socialism as, "more on the romantic side," with the essence of it being the freedom of the individual.[23] This allowed him to deflect the criticisms of the USSR and Stalinism away from himself. At the same time it provided a connection to the USSR he could not break. Stewart stated that he "trusted the Marxism of the Soviet Union to have the right answers for the advance of the international Many towards the fulfillment of all their capacities. But Russia was the only country of Marxism," and he did not think him, "could abandon Stalin without surrendering my life raft."[24] This belief is important to remember in understanding the dupe. Many found communism appealing and could not reject it outright. However, the most attractive quality in people was naiveté. This was another characteristic Stewart had. In describing himself and Hollywood, he stated:

> Hollywood itself was in its own orbit. "Reality" was whatever was recorded on film at the end of each day's shooting, and events of political or economic significance were left to the newsreels. It was no wonder that I was later to fit so successfully into this dream world, for I myself was living happily in a personal paradise in which no essential realities were allowed to disturb my determined play and pleasant dreams. I was indeed a lucky fool—on an extended leave of absence from infelicity."[25]

This mentality went a long way in helping the Soviets achieve their goals. Through filtering directions from Soviet agents through many levels the origin of the information was distanced from the source.

Front groups and unions were key tools and demonstrate how the Soviets tried to achieve the other goals they had for the film industry. In his autobiography, Ronald Reagan stated that the infiltration of Hollywood took place in a three-step plan. The first step was the creation of a massive strike in Hollywood, leading to the second of using the strike as premise for creating one large film union, that would then make it possible to move to the third of gradually working into movies the, "requisite propaganda attitudes to soften the American public's hardening attitude toward communism."[26] The first step can be seen in the actions of the Conference of Studio Unions and the attempts of its founder, Herb Sorrell, to control all the Hollywood unions by trying to get every union to strike if one chose to do so.[27] To actually take control of a union, they used the organizations' standing rules of only needing 10 percent attendance to hold a meeting to their advantage. Communists made sure their members attended while the other, less interested, members worked on meeting deadlines.[28] This allowed them to put into strategic places the necessary writers and editors to control the content of scripts.[29] This was the last step taken to influence movies. They now could influence film content.

When it came to content manipulation, the plan was never to really make a propaganda piece. The basic mission was to simply keep any anti-communist, Stalinist, or USSR content out of the films.[30] It was more important to do no harm rather than do good. The tactics used are best described through John Lawson, who stated that "as a writer, do not try to write an entire Communist picture, but try to get five minutes of Communist doctrine, five minutes of the party line in every script that you write."[31] Furthermore, this tactic even had protocols built in to keep doctrine in the script even if people recognized it as communist ideology. This was achieved through the careful placement of ideological statements in expensive scenes with the high-cost stars. The expense argument prevented it from being cut.[32] A general example of how propaganda was smuggled into films can be seen in *The Spirit of Culture*. The writers inserted the line, "it is far better to die on your feet than live as slaves on your knees." The line came directly from a communist leader in Spain.[33] It wouldn't be noticed by the average person, but it had great meaning for a communist. Another example is the film *Action in the North Atlantic*, written by John Lawson and starring Humphrey Bogart. The film portrayed the communist maritime union shipping Lend-Lease supplies to the USSR in a favorable light and the USSR itself as, "America's most noble and worthy ally."[34] However, the most overt example of using film as a propaganda tool was *Mission to Moscow*. The film attempted to justify Stalin's purge trials, the invasion of Finland, and develop the concept of the USSR as an ally.

## How It Ended

The front groups, mainly the Hollywood Anti-Nazi League, and people's individual actions proved to be the campaign's downfall. The most influential event was the signing of the Molotov-Ribbentrop Pact. The anti-fascist stance had been pushed so hard that when this move was announced, it was hard for communist sympathizers or anti-fascists to accept it. Donald Stewart wrote that, while the "pact with Germany was a hard pill to swallow," he continued to trust the USSR to have the "correct Marxist understanding of the situation."[35] For those loyal to communism, this was something that could be justified. But those who were truly anti-fascist could no longer overlook the group's communist connections. As a result, the group's membership fell.[36] In response to wavering support, League leaders tried to turn the focus away from the League's contradictions by claiming their new purpose was to maintain peace or that the pact was merely a move that allowed the USSR more time to build its defenses. Additionally, the talking points included criticisms of the efforts to supply the allies as wells as the attempts of the U.S. to develop its own military. These actions were labeled as attempts to drag the country into the war.[37]

Another blow that severely weakened this campaign came from the House Un-American Activities Committee (HUAC). Initially, HUAC was used to browbeat critics with claims that the government was attacking the people's First Amendment rights. However, the Committee's evidence was too compelling. People like Dorothy Parker had left a trail so easy to follow that is was impossible to defend her against claims that she was not a communist sympathizer. Her FBI file documented activities in other Munzenberg front organizations, and by the early 1940s Congress had enough evidence to convince "willing accomplices" or "dupes" into, "telling the truth about the subversive effects of communist influence in Hollywood."[38] The evidence and questions presented to people like Humphrey Bogart was so compelling that after his time before the Committee he felt compelled to issue a statement declaring his anti-communist position and later yelled, "You f—ers sold me out!" to those who had tricked him.[39]

## The Impact on American Society

The effects of this active measures campaign have been debated for decades, but one would have a difficult time suggesting that its effects have washed off. Kenneth Billingsley, author of *Hollywood Party*, points out that actors and actresses like Ed Asner and Jane Fonda have openly embraced communism without detriment to their careers. One might even be able to include Sean Penn in this list. Billingsley brings up the reenactment of the HUAC hearings in Hollywood in 1997. The event focused more on the perceived suppression of speech rather than the revelation that communists had indeed been operating in Hollywood.[40] Furthermore, the desire to suppress the crimes of communism may still be at work today as there is no shortage of films that depict events such as the atrocities in Rwanda during the 1990s, yet there are few, if any, Hollywood films that place into context the tens of millions of lives taken in the name of communism.

## Conclusion

The active measures campaign in Hollywood was a set of activities that exemplifies the overall strategy of Soviet operations. The Soviets were constantly on the offensive and made extensive use of their love of the lie. With patience, they employed many layers of subtleties and lies and half-truths to conceal involvement while exploiting every medium possible. The result was a successful infiltration of Hollywood whose effects are still felt today.

## Notes

1. Paul Kengor, *Dupes: How America's Adversaries Have Manipulated Progressives for a Century* (Wilmington, Del.: ISI Books, 2010), 182.

2. Ronald & Allis Radosh, *Red Star Over Hollywood: The Film Colony's Long Romance with the Left* (San Francisco: Encounter Books, 2005), 110.

3. Stephen Koch, *Double Lives: Spies and Writers in the Secret Soviet War of Ideas Against the West* (New York: Free Press, 1994), 5

4. W.G. Krivitsky, *In Stalin's Secret Service* (New York: Enigma Books, 2000), 54.

5. Kent Clizbe, *Willing Accomplices: How KGB Covert Influence Agents Created Political Correctness, Obama's Hate-America-First Political Platform, and Destroyed America* (Ashburn, VA: Andemca Publishing, 2011), Kindle version at location 4821.

6. Kenneth Billingsley, *Hollywood Party: How Communism Seduced the American Film Industry in the 1930s and 1940s* (Rocklin, CA: Forum, 1988), 48 & 80 & Kengor, *Dupes*, 182.

7. Koch, *Double Lives*, 91.

8. Koch, *Double Lives*, 90.

9. Allen Weinstein & Alexander Vassiliev. *The Haunted Wood: Soviet Espionage in America – The Stalin Era* (New York: Random House, 1999), 110-112.

10. Koch, *Double Lives*, 40.

11. Radosh, *Red Star*, 39 & Billingsley, *Hollywood Party*, 24-25 & 27.

12. Radosh, *Red Star*, 39-40.

13. Koch, *Double Lives*, 27.

14. Koch, *Double Lives*, 27.

15. Koch, *Double Lives*, 249-250.

16. Koch, *Double Lives*, 15.

17. Billingsley, *Hollywood Party*, 58.

18. Koch, *Double Lives*, 250-251.

19. Salka Viertel, *The Kindness of Strangers* (New York: Holt, Rinehart, and Winston, 1969), 211.

20. Viertel, *The Kindness of Strangers*, 211.

21. Kengor, *Dupes*, 190-192.

22. Kengor, *Dupes*, 227.

23. Donald Stewart, *By a Stroke of Luck: An Autobiography* (New York: Paddington Press, 1975), 216.

24. Stewart, *By a Stroke of Luck*, 247.

25. Stewart, *By a Stroke of Luck*, 153.

26. Ronald Reagan, *Where's the Rest of Me* (New York: Duell, Sloan and Pearce, 1965), 163.

27. Billingsley, *Hollywood Party*, 42.

28. Radosh, *Red Star*, 44 & Billingsley, *Hollywood Party*, 65.

29. Billingsley, *Hollywood Party*, 91-93.

30. Billingsley, *Hollywood Party*, 92.

31. Kengor, *Dupes*, 183.

32. Billingsley, *Hollywood Party*, 91.

33. Radosh, *Red Star*, 87.

34. Radosh, *Red Star*, 88-89.

35. Stewart, *By a Stroke of Luck*, 249.

36. Koch, *Double Lives*, 253 & Billingsley, *Hollywood Party*, 77-78.

37. Billingsley, *Hollywood Party*, 76 & 80.

38. Clizbe, *Willing Accomplices*, Kindle version at location 4900.

39. Kengor, *Dupes*, 227.

40. Billingsley, *Hollywood Party*, 282.

# Integrating Cultural Geography with Psychological Operations: Islamic Superstitions

Lennea Mueller

The United States has found its soldiers in foreign lands, where learning about the local and regional culture is the essence to winning the fight. The U.S. has adapted the last decade during the war on terrorism to finding ways to defeat an asymmetrical enemy and has made gains learning about and understanding the nature of the land. With the increase of cyber -attacks, drones, and other advanced weaponry, war is becoming less personal, letting it at time be fought without actually fighting. One of the greatest weapons, however, is too often overlooked and overshadowed by technological advances: the weapon of the mind.

Although the Obama Administration may call the war in Afghanistan and Iraq a victory, the threat of Islamic terrorism directed at the U.S. will not cease to exist anytime soon. The enemies are not only the terrorists but also the ideology they follow. The current strategy of decapitation is only a quick fix and resources must

***Lennea Mueller*** *graduated from Northern State University in 2010 in Aberdeen, SD where she studied Political Science and Criminal Justice. She graduated in May 2013 from The Institute of World Politics, with a Master's degree in National Security Affairs with a concentration in Intelligence. Ms. Mueller has interned at the Westminster Institute and the online consulting company, Wikistrat where her research areas included counterterrorism and international crime. She also has held internships at the State Department –Bureau of Political Military Affairs, South Dakota Senator Tim Johnson's office, and for the South Dakota State House Majority Leader. Ms. Mueller currently works as Social Media Analyst for a non-profit. Her future plans include working within the intelligence community.*

be allocated to fight at the core of the conflict. Psychological operations (PSYOP), today called Military Information Support Operations (MISO), as defined in the U.S. Joint Military Doctrine 3-13.2, are "operations to influence foreign audience's perceptions and subsequent behavior as part of approved programs in support of the United States' policy and military objectives."[1] Psychological operations are the missing link to winning the ideological war which continues to breed terrorism. A colonel in the U.S. Army said it best: "[I]f terrorism is an inherently psychological phenomenon, then it should stand to reason that psychological operations would and should be a primary method of attack or defense at the global planning level."[2]

Planning begins with mapping out the geography of battle. Geography has been the cause of conflict, cooperation, and will remain to be a source of relations between people, states, and nations. Most individuals think about geography in the confines of tangible elements, but culture is also part of geography. It is the nature of the territory, the personality of the landscape, and defines the character of those who occupy it. Many scholars, such as Samuel Huntington, have stated that future wars will be caused by a clash of cultures, as one fails to understand the other. Although many disagree with his theory, one cannot ignore the magnitude that cultural geography plays in war and conflicts; it always has and it always will.[3] When cultural geography is not understood, the likelihood of peace between nations is dismal. Much like the art of intelligence, diplomacy, and statecraft, there is also an art of mastering geography, which includes the intangible element of culture.

When one cultivates knowledge of culture, one discovers a variety of avenues which can be exploited in PSYOP. "Mind War," a piece on the psychology of victory and stated that the objective of PSYOP is to "map the minds of neutral and enemy individuals and then change them in accordance with US national interest."[4] These operations must be planned in advance and engaged at the strategic and tactical planning levels.[5] During these operations, the ethics and values of the community must be respected or the effects will not produce the anticipated results; therefore the entire spectrum of the target audience must be analyzed with the same vigor as the enemy. The enemy and its supporters must be psychologically engaged before they pick up arms, meaning that PSYOP can be appropriately waged before, after, and during conflict and in times of peace, war, and the grey are in between.[6] The art of psychological operations is to make the adversaries believe they are making their own decisions, which the operator will be influencing. However, they cannot be coerced; the U.S. must be different than the enemy and maintain the trust of the neutral audience.[7]

Islamic terrorism is a political and religious movement; its adherents kill because they believe they are directed by Allah. A PSYOP campaign that should be explored is to play on the religious fears, and superstitions of these devout Islamists. Spirituality is a central component of the "psychology of war and its actors."[8] Beginning with how they perceive victory.

Terrorist gage success by their own death and the death of others; using human life as both an offensive and defensive weapon; already they have an advantage. When facing an enemy like this, "all methods of deterrence and destruction can do nothing"[9] for the enemy has already turned death into a "counter-offensive weapon."[10] The question then is: How does one defeat an enemy that does not fear death, for death bring them paradise? These militants fight for a radical ideology with a plethora of superstitions rooted in a deep history of fears, providing an ideal target for a successful psychological operation because demagogic ideologies tend to be more vulnerable.[11]

Most successful terrorists are devout, young, single men that are recruited in religious institutions, outside mosques where individuals hand out flyers inviting potential recruits to information sessions about their movement, and even in schools. Recruiters also tend to have a penchant for individuals coming straight out of prison who lack homes, family, or financial resources.[12] The Internet has also proved to be invaluable tool for terrorists, not only for recruiting but to give out orders, disseminate bomb-making instructions, and issue fatwas, enabling them to reach a large and disperse audience.[13] Most terrorist organizations understand the importance of proper training and preparation psychologically when it comes to jihad,[14] so the recruitment process can begin as young as ten years. Madrassas are schools where many of the terrorists start their training. With free room and board, children are sent for religious education and taught the Koran, specific to Wahabi Islam since Saudi Arabia funds many of the madrassas regionally.[15] Recruits learn how to surrender themselves to the cause. Former terrorists have been interviewed stating they lived in perfect content in their ignorance, silently following orders from their spiritual leaders. This recruitment process is where a PSYOP campaign must be waged, striking the root of where this toxic belief system grows.

As terrorists are taught how to interpret the Koran, they are also exposed to the many superstitions that can be utilized in a psychological campaign. They can act as a deterrent for an offensive terrorist attack or explain the phenomena of why they lost or were destroyed in the event of an attack against them. The first superstition to explore is the text surrounding the presence of the "evil eye," found in the Koran: "The influence of an evil eye is a fact, if anything would precede the destiny it would be the influence of the evil eye."[16]

The evil eye can be source of bad luck, disease, envy, and jealously. Many Muslims believe that if it is used against them, they must remove the alleged effects of the evil eye by, for instance, gargling with water, then washing their body with that same water, and then finally tossing that same water behind their back.[17]

Devout Islamists are extremely suspicious of idols—both non-Muslim and Muslim—as well as of saints' tombs and shrines. This is evident from the current events unraveling in Mali where the Islamist group Ansar Dine, has vowed to destroy all mausoleums and shrines in the country stating they are "idolatrous" to the religion.[18] Saudi Arabia also famously destroyed many of its ancient tombs of saints.[19] Other idolatrous actions feared in Islam include: Koranic verses written on amulets, written backwards, or written in blood.[20] Additionally forbidden in the Koran is black magic, and some Muslim countries have arbitrarily arrested women suspected of practicing magic and voodoo.[21]

All the above superstitions could be incorporated into an effective PSYOP campaign; the most exploitable belief to use would be the paralyzing fear of jinn spirits and Satan. The name jinn come from *itjinan,* meaning "concealed."[22] The Koran says, "surely he sees you, he and his tribe, from where you see them not."[23] Jinn are not humans, nor are they angels; they can think and reflect and choose between good and evil. They can take three forms: flying in the air; traveling about in human form on land; and those which resemble dogs and snakes.[24] Satan, also known as Ilblis, is malicious jinn[25] who walks beside every person and is present during all affairs of man.[26] Although the human eye is too weak to see them, some believe jinn can actually take the form of humans,[27] even able to marry and have children. Muslims are very mindful about where they pray, since jinn can be found anywhere thought to be impure: abandoned settlements, bathrooms, sewage, garbage, open deserts, and fields.[28]

Jinn also can possess whatever and whomever they wish. In addition to possessing animals and trees, they can take over the human mind.[29] Women are believed to be more prone to possession, especially *illiterate* women.[30] In Somalia and Pakistan, some believe that less educated men are also more susceptible to possession and should be more fearful of jinn.[31] In addition, Iraq is believed to be a place of jinn, sorcery, and disease.[32] As a result, Muslims are very careful to avoid activities forbidden in the Koran that may leave them vulnerable to the power of jinn—such as drinking alcohol, gambling, or building statues and alters which could be used to take the place of worship of Allah.[33] Additionally, believers are told to enter bathrooms with the left foot first, spit to one's left side only, and use only the left hand when using the bathroom. Believers are urged to refrain yawning because it is a reflection of laziness, which Satan sees as a sign of weakness.[34] Many Muslims believe that Satan whispers and urinates in the ears of those sleeping who neglected to pray the night before and that if urine is found on clothing it would be means for punishment in the after-life.[35] They are taught that the plague was the "spear of Satan"[36] and a fever is the heat from hell.[37]

Not only do jinn walk with humans, but animals can be servants to jinn.[38] A black dog for example is the devil and the same is true for black cats because "black rallies satanic forces more readily than any other color."[39] Animals, unlike humans, can sense a jinn's presence and see what humans cannot.[40] Muslims are warned: "When you hear the cock crowing then ask Allah for his bounty for it has seen an angel. When you hear the braying of a donkey, and then seek refuge in Allah from Satan for it has seen Satan."[41]

Different regions believe different things, hence underscoring the need for cultural expertise. For example, scorpions are another animal that can serve jinn, but there is some disagreement among Muslim scholars as to if scorpions are the source of evil or can protect humans. In Africa, scorpions have been feared since ancient times because during summer months in rural areas, inhabitants were susceptible to scorpion bites while walking barefoot in the dark. While other regions are famous for their scorpion bead-work, worn not only to ward *off* scorpion bites but also to offer protection from the evil eye.[42]

PSYOP has used the exploitation of superstitions in past conflicts but none better illustrates the mastery of integrating cultural superstition into an effective campaign than one that took place in Philippines. In the 1950s, Air Force General Edward G. Landsdale had his psychological warfare squad plant stories in villages of vampire sightings in areas critical to the Philippine insurgents, exploiting the local culture's fear of vampires. Two nights after the rumors were allowed to resonate; he planned an ambush on the insurgents as they passed through the vampire-ridden area. His squad picked off one insurgent and administered two punctured wounds to his neck and drained the blood out before putting him back on the trail. When the insurgents went back looking for their lost man, they found the "vampire attack" and quickly left the region, leaving this strategic location open for General Landsdale's troops. As this one example demonstrates, "fear is a weapon just as surely as is a rifle or a tank; if you frighten your enemy enough, you may defeat him without having a fight."[43]

Exploiting the enemy's superstitions in a PSYOP campaign has been done more recently in Iraq during Operation Enduring Freedom. On December 8, 2001, U.S. soldiers dropped leaflets on Iraqi villages; one particular leaflet, *AFD 56b,* displayed a picture of Osama bin laden at the top, and a second picture of him below as if possessed by a snarling jinn.[44] The military also explored the possibility of using a holographic image of a flying Allah with messages to persuade Iraqis to cooperate. [45] These efforts are a move in the right direction, but the U.S. still lacks the political will to create and use unorthodox methods in psychological warfare.

Cultural superstitions cannot only be used tactically, but can also be integrated into a grander strategy toward combating terrorism. If a terrorist is convinced that his act is not for Allah but that he is being influenced by jinn, he not only humiliates himself, but also his family, village, and religion. Bin laden famously said that death did not scare him but humiliation did.[46] What would happen if terrorists were to hear rumors or had some sort of "evidence" to support accusations that they were taking spiritual guidance from a possessed imam (spiritual leader)? What would follow if they were manipulated into thinking their deaths would be a source of evil onto their families from an evil jinn? What would follow if damaging videos or pictures leaked providing proof that spiritual leaders were engaging in forbidden acts or manners? The possibilities are endless, but they call for innovation, imagination, and the willingness to try something new.

If terrorism is to be dealt with globally and strategically, the focus must be on what makes the terrorist a terrorist, striking at the source of this bad echo.[47] Scholar of jihadist ideology Patrick Sookhdeo suggests that the interpretation of jihad must be reformed in accordance with common values.[48] In order to achieve this, Islamic institutions, schools, madrassas, and any place that teaches jihad, must integrate the reformation.[49] "If an enemy is to be defined, then the enemy is not Muslim but the classical interpretation of Islam."[50] Terrorists believe they are killing and dying as a means to get to paradise, and this needs to change.[51] A suicide attack is only martyrdom if an Islamic leader states that it is, therefore an Islamic leader has the power to declare it not so.[52] There are many challenges with reforming the religion. It would be imperative that the West's participation be covert because if it became known, the whole reformation would be lost and might call for renewed violent jihad.[53]

Western cultures have found themselves victims of terrorism and must find new approaches in combating global terrorism. Because of both fiscal means in political and military affairs and American citizens' dwindling appetite for fighting wars abroad, PSYOP can be a low-cost, high-impact solution. It can be part of the ongoing struggle to maintain national security without the use of weapons. The U.S. should encourage innovative thinking within its national security apparatus to devise more advanced psychological warfare programs. If the U.S. is to win over the hearts and minds of those who have declared themselves America's enemies, or at least the multitudes of people living under their control, the U.S. must target what its enemies hold close to their hearts and disassemble the falsehoods which have been infused in their minds. Cultural geography integrated with psychological operations is the key to dismantling superstitions that leave so many minds vulnerable to believing just about anything.

## Notes

1 "Joint Publication 3-13.2: Psychological Operations". United States Military, January 7, 2010. http://www.fas.org/irp/doddir/dod/jp3-13-2.pdf, p.1

2 Boyd, Col. Curtis. "Army IO Is PSYOP: Influencing More with Less." *Military Review* (June 2007): 67–75, p. 21.

3 Huntington, Samuel. "Clash of Civilizations." *Foreign Affairs*, 1993.

4 Valley, Col. Paul E. *From PSYOP to MindWar: The Psychology of Victory*. San Francisco, California: Headquarters, 7th PSYOP Group US Army Reserve, 1980, p.4.

5 Ibid, p.5.

6 Ibid.

7 Ibid, p.9.

8 Wlodarczyk, Nathalie. *Magic and Warfare: Appearance and Reality in Contemporary African Conflict and Beyond*. Macmillan, 2009, p.30.

9 Baudrillad, Jean. "The Mind of Terrorism." *The European Graduate School*, n.d. http://www.egs.edu/faculty/jean-baudrillard/articles/the-mind-of-terrorism/.

10 Ibid.

11 Valley, p.1.

12 Sookhdeo, Patrick. *Understanding Islamist Terrorism: The Islamic Doctrine of War*. UK: Isaac Publishing, 2009, p.133.

13 Ibid, p.129.

14 Ibid, p.138.

15 Ibid, p.135.

16 "Myths of Muhammed." *ReligionofPeace.com*, 2011. http://www.thereligionofpeace.com/Muhammad/myths-mu-superstition.htm.

17 Ansari, Moiz. *Islam And the Paranormal: What Does Islam Says About the Supernatural in the Light of Qur'an, Sunnah And Hadith*. Universe, 2006, p.201.

18 "Islamists Vow to Smash Every Mausoleum in Timbuktu." *BBC*, July 1, 2012. http://www.bbc.co.uk/news/world-africa-18665522.

19 Stewart, Rory. *The Places in Between*. Houghton Mifflin Harcourt, 2004, p.80.

20 Philips, Abu Ameenah Bilal, trans. *Ibn Taymeeyah's Essay On the Jinn*. 3rd ed. International Islamic Publishing House, 1996, p.21.

21 "Myths of Muhammed."

22 Ashour, Mustafa. *The Jinn in the Qur'an and the Sunna*. Dar Al Taqwa, 1998, p.8.

23 "Al-Araf Ch. 7 Verse 27." In *Qu'ran*, n.d.

24 "The World of Jinn." *Religion of Islam*, August 8, 2010. http://www.islamreligion.com/articles/674/.

25 Kasem, Abul. "Islamic Voodoos: Angel and Jinn." *IslamWatch.org*, n.d. http://www.islam-watch.org/.

26 Ashqar, □Umar Sulaym□n. *The World of the Jinn and Devils*. Islamic Books, 1998, p.5.

27 Ashour, p.7.

28 Philips, p.33.

29 "Myths of Muhammed."

30 "Jinn: Born of Fire." *The Economist*, December 19, 2006. http://www.economist.com/node/8401289.

31 Ibid.

32 "Book 54 Ch. 54 Verse 11." In *Qu'ran*, n.d.

33 Ashqar, p.118.

34 Ashour, p.47.

35 El-Neil, Ibn. *The Truth About Islam*. Strategic Book Publishing, 2008, p.138.

36 Ashour, p. 47

37 El-Neil, p.138.

38 "Jinn: Born of Fire."

39 Philips, p.27.

40 Ashqar, p.13.

41 Ashour, p.38.

42 Frembgen, Jürgen Wasim Frembgen. "The Scorpion in Muslim Folklore." *Asian Folklore Studies;* 63, no. 1 (2004): 95, p.107.

43 Elliston, Jon. "Psywar Terror Tactics." *Parascope.com*, n.d. http://www.parascope.com/ds/1096/psy.htm.

44 Friedman, Herbert. "Superstition PSYOP." *Psywarrior.com*, n.d. http://www.psywarrior.com/SuperstitionPSYOP.html.

45 Ibid.

46 Waller, Michael. "Larry, Curly and Musab." *Los Angeles Times*, May 21, 2006. http://articles.latimes.com/2006/may/21/opinion/oe-waller21.

47 Sookhdeo, p.159.

48 Ibid.

49 Ibid, p.160.

50 Ibid, p.168.

51 Baudrillad.

52 Sookhdeo, p.157.

53 Ibid, p.161.

# The American Culture of War and COIN

Richard Fiesel

Despite having emerged as the world's preeminent military power with the most influential and widespread culture in world history, the United States has always struggled to accept and deal with the existence of war.[1] This struggle originates from the inherent inequality[2] and degradation of man in warfare.[3] While many nations and states have easily accepted these unavoidable characteristics of war, Americans have historically found these qualities to be in direct conflict with the founding principles of their nation and have developed a doctrine and culture of war that attempts to preserve these principles. When this culturally acceptable mode of war is not adhered to, however, American public opinion often turns against the war, the Administration, or, at times, the United States itself.[4]

The two most influential and foundational tenants of American society are the beliefs in the equality of man and the charge that mankind is not a means to an end for a society, but rather the end itself. The belief that man is the end of the nation in particular has dictated how America wages war. These two ideas were pushed to the forefront of American identity with the signing of Declaration of

***Richard Fiesel*** *earned an M.A. in Statecraft and National Security Affairs with a specialization in Intelligence in May 2013. Mr. Fiesel is currently a Researcher for the Asia-Pacific and Counterterrorism Desks at Wikistrat Global Consulting, where he also serves as an analyst for company and client geostrategic simulations/war-games. In conjunction to these roles, Mr. Fiesel also serves as a recruiter for Wikistrat, maintaining correspondence with prospective employees and universities. He is currently seeking post-academic work where he can continue to hone and utilize his skills and experience in counter-terrorism and geostrategic analysis.*

Independence, a document that has arguably served as the universal moral code of American society. Together they explain America's reluctance to enter war, its tenacity in the conduct war, the people's repulsion to casualties, its empathy to civilians, and, consequently, the nation's perceived obsession with military technology.

Just as the social and political realms of American society toiled to harmonize these ideas in the arenas of suffrage and racism, American military leaders have also wrestled with the challenge of infusing these principles into the structural and operational doctrine of the United States military.[5] Despite Americans' best efforts, however, the American way of war is by no means invincible and the U.S. is constantly attempting to fit lessons learned into its model of war without trampling on the tenants of its culture.

For instance, after the Korean War, the United States finally came to grips with the reality that the world required it to maintain a competent professional military despite cultural reservations against an institutionalized military. Over time, American culture was able to accept this reality into its cultural perception and the professionalism of the U.S. military is a point of pride among most Americans today. In contrast, the ominous presence of Vietnam on the American conscience remains so enduring due to the incapability of America to acknowledge and adapt to the unconventional realities of the conflict. In fact, the sole reflective product in the immediate post-Vietnam era is promulgated in the anti-interventionist sentiment crystallized in the Powell Doctrine promoted by the former chairman of the Joint Chiefs of Staff, General Colin Powell. Simply put,

> Powell cautioned against intervention unless (1) the objectives were both vital to US interests and very clear, (2) the US could bring overwhelming force to bear in the conflict, (3) Congressional and public support could be assured, and (4) a clear exit strategy was in place.[6]

The problem with the Powell Doctrine as the sole lesson of the Vietnam War lays in the fact that it was only a political, not a cultural, or even military, change. As a result, its principles were likely to be adhered to only so long as its advocates remained in political power. Even then its core tenants are so specific and limiting that the realities of the political or international arenas often times forego any possibility to appease them. Indeed this rendition of the wisdom gained from the conflict provided guidance on when the U.S. should intervene—an important consideration affirmed in a joint Army/Marine Corps Field Manual (FM 3-24)—but had little to say about what would happen if the U.S. again found itself immersed in a Vietnam-like counterinsurgency (COIN) campaign despite caution on the part of military leaders.[7]

With the end of the Iraq War in 2011, America once again worked to digest the lessons of the hard fought conflict. The cornerstone of this reflection has been the FM 3-24 and the man behind it: General David Petraeus. In the manual's foreword, John A. Nagl of the Center for a New American Security writes that "the story of how the Army found itself less than ready to fight an insurgency goes back to the Army's unwillingness to internalize and building upon the lessons of Vietnam... After the Vietnam War, we purged ourselves of everything that had to do with irregular warfare or insurgency because it had to do with how we lost that war."[8]

Recognizing the mistakes made not only in Vietnam itself but also in the American reflection of the war led to a reimagining of the conflict that focused on a full spectrum (political, military, social), doctrinal review. It credits a challenging, flexible, and resilient Vietnamese foe for America's defeat and attempts to learn from that foe. The very fact that the dominant document to arise from this contemporary examination was an Army/Marine Corps field manual points to the tactical and operational lessons the reflection on Iraq borrowed from Vietnam. This reflection stresses that the U.S. could have won the war had it adopted its successful "clear, hold, and build" tactics earlier in the war, or had U.S. public support held out post-1968 when the war was actually being won.[9]

Petraeus saw the American tendency to ignore counterinsurgency as a fatal flaw in the American culture of war. He explained that "American involvement in counterinsurgencies is almost universally regarded as more likely than in most other types of combat—more likely, for example, than involvement in high intensity conflict on the plains of NATO's Central Region."[10] Even so, he saw the paradox this created for military leaders.

> The senior military thus finds themselves in a dilemma. The lessons taken from Vietnam would indicate that, in general, involvement in a counterinsurgency should be avoided. But prudent preparations for a likely contingency (and a general inclination against limiting a president's options) lead the military to recognize that significant emphasis should be given to counterinsurgency forces, equipment and doctrine. Military leaders are thereby in the difficult position of arguing for the creation of more forces suitable for such conflicts while simultaneously realizing they may advice against the use of those forces unless very specific circumstances hold.[11]

In many ways, the FM 3-24 offers a belated solution to the problems faced by Westmoreland and Abrams in Vietnam and further embedded lessons that emphasized how counterinsurgency could be done correctly, rather than those which questioned whether it should be done at all. America has traditionally suffered from mirror-imaging and in many ways has assumed that the morality and nobility of its cause—the rightness of America's political decision to intervene as described by the Powell Doctrine—and its doctrinal adherence to international law and human civility would be enough to prevent the rise of, or even defeat, an insurgency. History has shown otherwise.

While the FM 3-24 and Petraeus have certainly made an impact in both political and military circles, their effect on the perception of war among the American citizenry remains to be seen. Historically, the limited and abstract natures of U.S. objectives in counterinsurgency operations have been difficult for most Americans to accept. Due to the immense value they place on life, Americans have struggled to accept waging war other than one characterized as defensive in intent, offensive in nature, tangible in its course and effects, and total in means.

America's cultural repulsion to limited war stems from its implied limit on the use of firepower, technology, and other life-saving apparatuses integrated into conventional U.S. doctrine. The particular paradox of necessarily using the minimum force required, rather than maximum force available in order to succeed in COIN operations and save U.S. military and civilian lives in the long term, runs contrary to almost all that the U.S. culture of war adheres to. In many people's minds the very reason why the U.S. taxpayer spends billions of dollars on new and increasingly destructive weaponry is to prevent American loss of life and to keep American soldiers out of harm's way as much as possible.

Furthermore, COIN is inherently a defensive endeavor. While necessary for victory, the implications of prioritizing the protection of the population runs contrary to the U.S. perception of offensive war as the only means to victory. Protracted defensive conflicts are difficult for Americans to stomach as continued casualties are inflicted upon their forces while no land or other tangibles are gained.

It is also difficult for many Americans to grasp how COIN efforts affect their personal safety. Whether it's the preventative effort attempting to deny Afghanistan as a base of operations for terrorists, or the presence of a functional and stable democratic Iraq in the Middle East as a flashpoint for political change, the effects of these operations on an American's personal security are often of the second, third, or even fourth degree and are predicated upon many "what ifs." For better or worse, Americans tend to respond only to immediate and tangible consequences and typically dismiss long-term opportunities and securities as too removed from the present to influence or plan for. In a way, American culture may too pragmatic for its own good.

Though these cultural leanings and expectations may reside in the individual, they affect the sustainability and perceived legitimacy of COIN operations and foreign interventions. Realistic expectations regarding the longevity of the conflict, its degree of success in achieving national aims, and the type of war to be fought can help a population serve as a bastion of resolve for its armed forces and increase the chances of national success in a theatre. While political and military leaders play a significant role in crafting these expectations and should stress realistic objectives and expectations regarding any conflict, the citizenry must alter its views of success and victory to meet the realities of these conflicts rather than having political and military objectives attempt to meet the lofty idealism of the citizenry.

Luckily there are signs that the re-imagination of the Vietnam War, the experience in Iraq, and subsequent doctrinal change among America's policy and military professionals are leading to popular cultural change in America's view of war. Though it is uncertain if these changes will stand the test of time, it remains certain that America will never be able to consistently sustain effective overseas COIN operations unless U.S. aversion to limited and abstract war is soothed. Americans must realize that COIN operations and limited war is not a science, but an art. If the U.S. citizenry can internalize the strategic and operational lessons of these two conflicts while avoiding the selective amnesia that marked post-Vietnam America, the U.S. could truly reverse its historic political and social apprehension in combating insurgencies. So long as guerilla warfare remains an American weakness it will continue to characterize American warfare. In the dynamic game of evolution and counter-evolution that characterizes warfare it is America's turn to evolve.

## Notes

1 Adrian R Lewis, *The American Culture of War: The History of U.S. Military Force from World War II to Operation Iraqi Freedom* (New York: Taylor and Francis Group, 2007), 34-35

2 Ibid., 22.

3 Ibid., 34.

4 Ibid., 6.

5 Walter Millis, *Arms and Men* (New York: G.P. Putnam's Sons, 1956), 39-40.

6 Fitzgerald, "Rebirth of Counterinsurgency," 152.

7 Ibid.

8 Army and Marine Corps, *Counterinsurgency Field Manual*, Kindle edition, 155.

9 Bankus, "Learning from the Past," 10.

10 David H. Petraeus, "The American Military and Lessons of Vietnam: A Study of Military Influence and Use of Force in the Post-Vietnam Era," unpublished PhD thesis, Princeton, New Jersey, 1987, 307.

11 Ibid., 308.

# Euro-Army Reveries and Transatlantic Realities

Karina Rollins

The European Union, best known for its common market and single currency, also has a shared foreign and defense policy. While this policy is EU-specific, the idea of a common European defense is not. Modern European nations have formed military alliances with each other long before the creation of the European Union in 1992.

Some intra-European alliances formed shortly after World War II were intended as a deterrent to German aggression, such as the Treaty of Dunkirk between France and Great Britain in 1947. In 1948, Belgium, Holland, and Luxembourg joined the alliance by signing the Treaty of Brussels,[1] which led to establishment of the Western Union Defense Organization that same year.

In 1949, ten Western European countries joined the United States and Canada in creating NATO, the North Atlantic Treaty Organization, as protection against the anticipated Soviet alliance with Eastern European nations.[2]

## Dreams of a European Army

Even the idea of a European military is not new. Just one year after the creation of NATO, French Prime Minister René Pleven introduced a plan calling for a European Defense Community among six nations: France, Italy, West Germany, and the Benelux countries. Pleven had intended his plan as an alternative to West

***Karina Rollins*** *is a research editor at the Heritage Foundation, and is enrolled in the Statecraft and National Security Affairs program at The Institute of World Politics..*

Germany's proposed accession to NATO—and "to prevent instant German rearmament..."[3] (The plan was ultimately rejected by the French National Assembly, and West Germany joined NATO in 1955.)

In 1952, the same six countries signed a European Defense Community Treaty, "which provided for the integration of the armed forces of [these six countries] into a single European army."[4] A European Defense Community (EDC), however, was never established, due to Gaullist opposition in the French parliament. In 1954, the original 1948 Treaty of Brussels was amended to include West Germany and Italy; and the Western Union Defence Organization was renamed the Western European Union—intended as a replacement for the failed EDC attempt.

In 1961, French President Charles de Gaulle proposed the Fouchet Plan, with the goal of increased cooperation on defense, foreign policy, and economic matters among European countries. That plan, too, never took effect, largely due to opposition from Belgium, Holland, and Luxembourg, who worried that de Gaulle's plan would conflict with NATO. De Gaulle, believing that NATO was too heavily dominated by the United States and Great Britain, withdrew from NATO's integrated command structure in 1966, "arguing he had to preserve French independence in world affairs"[5] (though France did remain a formal member of the alliance).

## Cold War Developments

At the same time that European alliances—embedded in NATO under U.S. leadership—were aimed at resisting Soviet expansion, and despite the threat of nuclear attack posed by the Soviet Union, the 1980s were also a time of extensive anti-American sentiment in West Germany. Massive crowds protested the American Pershing II missiles that had already been stationed in West Germany in response to Soviet SS-20 missiles in Eastern Europe and the USSR.[6]

In September 1987, after a highly publicized Franco–West German military maneuver ("Bold Sparrow"), French President François Mitterand and German Chancellor Helmut Kohl launched an initiative to create a European Defense Council.[7] One month later, the seven-nation Western European Union (WEU) adopted a "defense policy platform" declaring the WEU's intention of developing a more "cohesive defense identity."[8] "Participating foreign ministers emphasized that their aim was to strengthen the West European pillar of the Atlantic Alliance, not to put into question Western Europe's enduring security partnership with the United States."[9]

## Paving the Way for a Common European Defense

The Treaty of Maastricht, which created the European Union in February 1992, specified a "common foreign and security policy" (CFSP) as one of the EU's three main components.[10] The treaty "committed EU member states to develop a foreign policy and, for the first time, allowed the EU to speak with a single voice in this area."[11] Then came Eurocorps, the "[e]mbryo of the European Army whose creation was decided on by F. Mitterand and H. Kohl in May 1992."[12] None of these developments was formally intended to create separation from NATO. In fact, at the NATO ministerial meeting in Berlin in June 1996, the Atlantic Alliance pledged to support the creation of a European Security and Defence Identity (ESDI) within NATO.[13] Eurocorps calls itself "a force for NATO and the European Union" as well as "a force for the European Union and the Atlantic Alliance."[14]

In 1996, Germany, France, Italy, and the U.K. established the Organisation for Joint Armament Cooperation (Belgium and Spain joined later). In 1999, the European Council "committed the EU to develop the ability to take independent military action . . . under the European Security and Defense Policy (ESDP)."[15] In 2003, the European Council adopted a European Security Strategy (ESS),[16] and in 2004, the EU created the European Defense Agency "to improve European defence capabilities in the field of crisis management and to sustain the [ESDP] as it stands now and develops in the future."[17] That same year, the European Constitution reiterated that the EU is charged with defining a common foreign and security policy.

Despite this dizzying array of EU defense declarations, the ESS states that the "transatlantic relationship is irreplaceable."[18] In 2009, President Nicolas Sarkozy announced that, after 43 years, France would rejoin NATO as a full member. "Sarkozy declared that rejoining the U.S.-led integrated command in Brussels will not diminish the independence of France's nuclear-equipped military and, on the contrary, will open the way for more French influence in deciding what NATO's new missions should be after the Cold War."[19]

Due to the myriad and contradictory developments in the EU, as well as often-vague declarations, it is not clear just what kind of an alliance a common European defense would be; nor is it easy to make sense of the military ambitions that would justify it.

## Collective Security Alliances

Writing on collective security during the Cold War, G. F. Hudson stated that the Warsaw Pact was "based on a community of political faith, professed by ruling parties which have long been accustomed to accept directives from a single centre, whereas NATO is an association of democratic states brought together by a sense of common danger, but without uniformity of doctrine or social pattern."[20] Given the growing trend in EU countries toward a unique "Euro" form of government and social doctrine, an EU defense amalgamation might share the "political faith" element of the Warsaw Pact. It certainly has an element of NATO (the association of democratic states)—but does it have a "sense of common danger" that differs from NATO's? Who is the shared enemy that a European defense force is intended to resist separately from NATO?

As Ambassador Robert Strausz-Hupé and Stefan T. Possony wrote in their examination of alliances, "a defensive alliance which has been concluded before the opposite camp has forged an offensive alliance of its own, or clearly manifested its aggressiveness, is a contradiction in terms."[21] What is the opposite camp to a European defense force? Why the insistence on a force that duplicates many NATO functions? Moreover, building up an EU defense would come at the expense of manpower and materiel that would otherwise supply NATO. Given talk by some EU officials in recent years of the European Union acting as a counterbalance to the United States,[22] the question arises whether the EU sees the U.S. as an unwanted hegemon.

## Money Matters

A very practical problem in building a European defense force is that "the global economic crisis has forced most European governments to trim their defense budgets: Germany will reduce defense spending by a quarter over the next four years, Britain's defense budget will be slashed by more than 8 percent in real terms by 2015, and the defense budgets of some of the smaller European nations have taken even larger cuts."[23]

These budget cuts are especially noteworthy considering that defense spending by European NATO members has already declined by 20 percent over the past two decades.[24] As Strausz-Hupé and Possony caution, "Alliances which do not assemble *superior* military power, or which do so only on paper because some of the allies have no readily available military strength, cannot be counted upon to avert war. This guarantee is given only by alliances which assemble *overwhelming* force."[25] For years now, as other analysts point out, "some European allies have only been able to afford modest contributions to NATO missions—even though they have supported the deployments politically. The Libya operation was the most recent illustration of this trend. As a result of the economic crisis, this 'participation' imbalance risks becoming more pronounced. Several European countries have already started withdrawing their troops from multinational operations in order to save money."[26]

Nevertheless, EU officials maintain that the European Union has already created viable military operations. While there is disagreement about the success of recent EU military endeavors, such as Operation Concordia in Macedonia in 2003 or EUFOR RD in the Congo in 2006, "the militarization of the European Union marks one of the biggest geopolitical shifts in the transatlantic alliance since the end of the Second World War."[27]

## Troubled Alliances?

While sharing transatlantic security challenges is certainly in the interest of all NATO members, especially that of the United States, which bears the greatest burden, there are warnings that the "U.S. should not confuse its desire to see European countries take on more security and defense responsibilities, both in Europe and in the wider world, with the ramifications of further European military integration—especially in terms of America's ability to build alliances."[28] Building alliances has become "increasingly problematic for Washington under the ESDP. Turkey's membership in NATO and Greece's and Cyprus's memberships in the EU present a profound conflict for the two organizations."[29]

France has already made several attempts at establishing a European defense identity for the purpose of countering American *hyperpuissance*, the term coined by former French foreign minister Hubert Védrine to criticize U.S. "hyperpower." In her 2002 book *Statecraft*, Margaret Thatcher stated that "France has for many years wanted to see an alternative military power to an American-led NATO. The European Union's plans for a separate integrated European defence provided the French with a unique opportunity to achieve this goal."[30] Central and Eastern European countries, on the other hand, "have long worried that divisions created by the ESDP might lead America to abandon its interests"[31] in Europe.

Despite this apparent threat to American power, the U.S. State Department and successive U.S. administrations have had a favorable impression of the ESDP. During the Berlin-Plus arrangements in 2003, the United States even agreed to ensure EU access to NATO operational planning, in addition to use of NATO assets and capabilities. In a major development, "Berlin-Plus also ensure[d] the adaptation of the NATO defense planning system to facilitate the availability of forces for EU operations."[32]

The EU, then, "has managed to negotiate the best of both worlds—a supranational public policy independent of American influence that is at least partly funded by America."[33] (Even de Gaulle's Fouchet plan had been intended to create a level of European unity "without supranational institutions."[34])

## European Defense—Part of the EU's Grand Strategy?

Given the duplication and draining of NATO missions and resources, various alliance conflicts, and significant voices in the EU that are hostile to the *status quo* of the U.S. as the world's sole super power, what then, is the function of the European Security and Defense Policy (now renamed the Common Defense and Security Policy)?

It is easy to point to anti-Americanism as the driving power behind a European defense force, and that phenomenon has played a role at times. But the situation is more complex: The long history of security alliances in Europe, large and small, includes many which have no evidence of considering U.S. influence one way or the other. It is not necessarily problematic to American interests that a continent with economic unity and a converging political identity should also consider a common defense.

What could be problematic for the U.S., however, is that the EU's Common Defense and Security Policy, so far, lacks a clear and cohesive mission, and comes at the expense of the existing transatlantic security ties—which could very well decrease, not enhance, security and stability in the Western world. Do a common market and political convergences constitute a basis for a supranational foreign and defense policy? Or, is foreign policy "an attribute of statehood that must remain at the nation-state level if it is to be meaningful or effective"?[35]

Some high-profile proponents of an expanding EU have talked openly of their hostility, not only toward the U.S., but also toward a worldview based on national sovereignty and democracy, embraced by the U.S. and its staunchest allies. Enrique Barón, for instance, president of the European Parliament from 1989 to 1992 who later joined the EU Committee on Foreign Affairs, Security, and Defence Policy, declared (referring to the Cold War) that the world order "provided by North American tutelage in a bipolar world of two superpowers" was "*unjust* if stable."[36] (Emphasis added.)

Like several EU proponents, Barón sounds as if he might have a political mission, not just against Ronald Reagan's America, but against Thatcher's Britain, and against Euroskeptics generally, claiming that their "opposition [to the EU] derives in part from the fact that the European Union stands as a more humane alternative to the low wage, free-for-all which, in Britain under Margaret Thatcher, spread insecurity as it brought down industries that with a little help might have survived."[37]

## Strategic Confusion?

Despite such declarations, EU officials take pains in their formal rhetoric to emphasize the importance of Euro-U.S. friendship, and the United States and NATO support many aspects of the EU's military ambitions—making the entire situation incoherent. Is the continued public emphasis on transatlantic cooperation mere lip service? With ever-evolving policies, ever-changing names for those policies, contradictory statements, ill-defined goals, and—despite all the talk of an autonomous EU military, very little independent EU military action—that question may not yet be answerable.

Why the continued insistence on building an autonomous European defense? As Sven Biscop and Jo Coelmont, authors of *Europe, Strategy, and Armed Forces*, argue, "the EU, the state-like actor that is the evident political expression of Europe, is developing into, and indeed *must* be, a strategic actor.... [I]n order to preserve its status as an economic power, the EU has no choice but to become a power across the board. That requires a grand strategy, and the means and the will to proactively pursue it."[38]

## A Common Defense—and a Grand Strategy—Must Be Defined

The European Union's defense alliance, then, is unlike other historic alliances. Its nature is not entirely clear; its goals are foggy, and any near-term funding for the undertaking seems fanciful. And, if the true goal is an actual European army (a coat of arms already exists[39]) with troops commanded not by military leaders from sovereign nations, but by supranational leaders, a European defense force would not be an alliance, but a single actor. It is not clear whether that actor would help to maintain a global balance of power, or whether it would destabilize the current Western security structure—possibly the last line of defense against the West's common enemies, from the Iranian regime to Islamist terror groups around the world, to increasingly anti-Western countries such as Venezuela and Russia.

EU grand-strategy proponents are focused on a "multipolar world," and their talk of transatlantic partnership often seems to emphasize "equality" for its own sake, such as in declarations that "a strong partnership can only be a partnership of equals"[40]—which (1) seems to imply a rivalry with the U.S. as the world's superpower, and (2) seems unrealistic when much of current European defense is still funded with U.S. tax money. Such realities, and the inconsistent developments on the part of the EU, have made it difficult to accept unquestioningly the claim that "European cooperation is not in competition with transatlantic cooperation..."[41]

It seems, then, that the time is now ripe for the United States, NATO leaders, and individual European nations to push the EU leadership for clarity and transparency as to just what the European Union's grand strategy entails.

## Notes

1. Officially named the Treaty on Economic, Social and Cultural Collaboration and Collective Self-Defence.

2. The ten European countries were: Belgium, Denmark, France, Iceland, Italy, Luxembourg, the Netherlands, Norway, Portugal, and the U.K.

3. Martin J. Dedman, *The Origins and Development of the European Union 1945–2008: A History of European Integration*, 2nd ed. (New York: Routledge, 2010), p. 71.

4. Enrique Barón, *Europe at the Dawn of the Millennium* (London: MacMillan Press, 1997), p. 205.

5. Edward Cody, "After 43 Years, France to Rejoin NATO as Full Member," *The Washington Post*, March 12, 2009, http://www.washingtonpost.com/wp-dyn/content/article/2009/03/11/AR2009031100547.html (accessed November 9, 2012).

6. As John Corry pointed out in 1983: "The paradox is that West Germany, under Chancellor Helmut Schmidt in 1977, was the first to propose that NATO mount a deterrence" to the Soviet missiles. Corry, "TV: Pershing II and West Germany," *The New York Times*, December 28, 1983.

7. Michael R. Lucas, *The Western Alliance After INF: Redefining U.S. Policy Toward Europe and the Soviet Union* (Boulder, Colo.: Lynne Rienner Publishers, 1990), p. 14.

8. David Buchan, "WEU Ministers to Set Out Charter on East-West Security," *The Financial Times*, October 26, 1987.

9. Lucas, *The Western Alliance After INF*, p. 14.

10. The Treaty on European Union, February 7, 1992, http://www.eurotreaties.com/lisbontext.pdf (accessed November 5, 2012).

11. Civitas, "EU Facts," April 2, 2012, http://www.civitas.org.uk/eufacts/FSEXR/EX3.htm (accessed November 5, 2012).

12. Barón, *Europe at the Dawn of the Millennium*, p. 204.

13. European Security and Defence Assembly/Assembly of Western European Union, "The EU-NATO Berlin Plus Agreements, "*Assembly Fact Sheet* No. 14, November 2009, http://www.shape.nato.int/resources/4/documents/14E_Fact_Sheet_Berlin_Plus[1].pdf (accessed November 8, 2012).

14. Eurocorps 1992–2012, http://www.eurocorps.org/operation.php (accessed November 8, 2012).

15. Civitas, "EU Facts." The ESDP is part of the common foreign and security policy (CFSP).

16. Javier Solana, "Foreword," in "European Security Strategy: A Secure Europe in a Better

World," Council of the European Union, 2009.

17. European Defence Agency, "Mission and Functions," 2012, http://www.eda.europa.eu/aboutus/whatwedo/missionandfunctions (accessed November 5, 2012).

18. European Security Strategy, "A Secure Europe in a Better World," Brussels, December 12, 2003, p. 13, http://www.consilium.europa.eu/uedocs/cmsUpload/78367.pdf (accessed November 9, 2012).

19. Cody, "After 43 Years, France to Rejoin NATO as Full Member."

20. G. F. Hudson, "Collective Security and Military Alliances," in Herbert Butterfield and Martin Wight, eds., *Diplomatic Investigations: Essays in the Theory of International Politics* (Cambridge, Mass.: Harvard University Press, 1968).

21. Robert Strausz-Hupé and Stefan T. Possony, *International Relations in the Age of the Conflict Between Democracy and Dictatorship*, (New York: McGraw-Hill, 1950), p. 229.

22. See, for example, Paul Reuber and Anke Strüver, *Politische Geographien Europas: Annäherungen an ein umstrittenes Konstrukt*, (Münster, Germany: LIT Verlag, 2005), p. 95; "Great Expectations: The Future of the European Union," Uncommon Knowledge with Peter Robinson, October 25, 2004, http://www.hoover.org/multimedia/uncommon-knowledge/27050 (accessed November 8, 2012); and Robert Kagan, "Power and Weakness," *Policy Review*, Vol. 113 (2002), pp. 3–28.

23. F. Stephen Larrabee, et al., "NATO and the Challenges of Austerity," RAND National Defense Research Institute, 2012, p. 1.

24. Anders Fogh Rasmussen, "NATO After Libya," *Foreign Affairs*, Vol. 90, No. 4 (July/August 2011).

25. Strausz-Hupé and Possony, *International Relations in the Age of the Conflict Between Democracy and Dictatorship*, p. 232.

26. Clara Marina O'Donnell, ed., "The Implications of Military Spending Cuts for NATO's Largest Members," Center on the United States and Europe at Brookings *Analysis Paper* July 2012, p. 8.

27. Sally McNamara, "The European Security and Defense Policy: A Challenge to the Transatlantic Security Alliance," Heritage Foundation *Backgrounder* No. 2053, July 18, 2007.

28. Ibid.

29. Ibid.

30. Margaret Thatcher, *Statecraft: Strategies for a Changing World* (New York: HarperCollins, 2002), p. 354.

31. McNamara, "The European Security and Defense Policy."

32. Ibid.

33. Ibid.

34. Andrew Moravcsik, *The Choice for Europe: Social Purpose and State Power from Messina to Maastricht* (Ithaca, New York: Cornell University Press, 1998).

35. Sally McNamara, "The EU's Common Foreign and Security Policy: How it Threatens Transatlantic Security," Heritage Foundation *Backgrounder* No. 2250, March 17, 2009.

36. Barón, *Europe at the Dawn of the Millennium*, p. xxiii.

37. Ibid., p. xi.

38. Sven Biscop and Jo Coelmont, *Europe, Strategy, and Armed Forces: The Making of a Distinctive Power* (New York: Routledge, 2012), p. 1.

39. Abhijit Pandya, "Do the Operations of the European Defence Force Have Democratic Legitimacy?" *The Daily Mail*, April 5, 2012.

40. Biscop and Coelmont, *Europe, Strategy, and Armed Forces*, p. 3.

41. http://www.iwp.edu/news_publications/detail/passion-for-cultural-diplomacy-inspires-german-iwp-studentIbid.

# The Emerging Threat to the United States Space Program

Kevin Cyron

"For The United States, the leading space faring nation for nearly half a century, to be without carriage to low Earth orbit and with no human exploration capability to go beyond Earth orbit for an indeterminate time into the future, destines our nation to become one of second or even third rate stature."

—*Neil Armstrong, James Lovell, and Gene Cernan in a 2010 open letter to President Obama*

There is a serious risk to the United States space program due to the indifference that several Congresses have shown the program. This indifference can drastically affect not only the progress in American science and technology but also American national security and prestige. As Steven M. Palazzo (R-MS), Chairman of the House Subcommittee on Space and Aeronautics stated, "Aero-

**Kevin Cyron** *graduated in May 2013 with an M.A. in Statecraft and National Security Affairs with a double focus in Public Diplomacy and Political Warfare and National Security and Defense Studies. He received an M.A. in Sociology and European Studies from St. Petersburg State University in Russia in 2008 and a B.A. in Communication from Marymount University in 2003. From 2005-2008, while attending school, he worked in St. Petersburg Russia for the American Chamber of Commerce and the Committee for Transport and Transit Policy for the City of St. Petersburg. In 2004, Mr. Cyron worked for the OSCE in Washington, D.C. before working for the European Parliament in Brussels. In Brussels, he worked for the Foreign Affairs Committee for the European Liberal Democratic Reform Party (ELDR) under Graham Watson (MP). From 1999-2003, while attending school, Mr. Cyron worked on Capitol Hill for Congressman Steny Hoyer from Maryland in his Congressional Office and also on his Committee Staff for the House Administration Committee. He currently teaches Social Studies and Religion at St. Luke School in McLean, Virginia.*

nautics research and development, and the technologies they spin off are critical to our national security, and to the ongoing success of our nation's aerospace industrial base, which is our country's greatest source of exports. No other enterprise has played a greater role producing innovative aeronautics technologies than NASA."[1] Furthermore, the National Security Space Strategy states: "Space capabilities provide the United States and our allies unprecedented advantages in national decision-making, military operations, and homeland security. Space systems provide national security decision-makers with unfettered global access and create a decision advantage by enabling a rapid and tailored response to global challenges."[2]

Since July 20, 1969, the United States has had a superior presence in space. This has not only helped the U.S. through the advancement of scientific knowledge, but has also given the U.S. a significant technological edge that has in turn provided military dominance for the U.S. and a higher standard of living for American citizens. The discoveries provided by the programs have led to this modern age as evident by modern communication via cell phones, workforce efficiency through computers, medical advancements that lead to better medicines, and countless others.

However, this American dominance of a presence in space is at risk due not only to the rising influence of China, but also from a former competitor, the Russian Federation, formerly the Soviet Union, and the fact that the U.S. lacks a coherent space policy. As Representative Palazzo stated, "The growth of overseas competition occurs at a time when NASA's aeronautics R&D funding is on the decline and continues to shrink..."[3] U.S. superiority in space can be sustained in the short term (five to 10 years), but not in the longer term. With NASA's budget constantly under political stress and with further financial cuts being made, this situation is likely to eventually lead to another significant Sputnik moment.

The ability of the U.S. to continue to have not only a presence, but also dominance in space is vital to the U.S. national interest, security, and national identity. A solid space policy is important, first, because through that policy, it allows the U.S. a scientific edge in knowledge about the Earth and our place in the universe. Second, there is the technology that is created as a result of such endeavors. Third, there are the defensive capabilities that can result in space research, such as a missile defense system and intelligence gathering devices such as satellites. Fourth, there is the economic aspect to space investment, in which hundreds of thousands of people can find meaningful employment through public and private enterprises. Fifth, it is quite simply the next step, and Americans should not stop after having made so much progress for the betterment of human civilization. As John F. Kennedy stated in 1962,

> We choose to go to the moon. We choose to go to the moon in this decade and do the other things, not because they are easy, but because they are hard, because that goal will serve to organize and measure the best of our energies and skills, because that challenge is one that we are willing to accept, one we are unwilling to postpone, and one which we intend to win, and the others, too.[4]

President Kennedy correctly points out that the challenges of space are ones "we are unwilling to postpone." Yet, this is exactly what the U.S. is doing and has been doing for the last several decades, leaving it to other rising powers such as China and Russia.

Starting with major budget cuts and cancelations of space programs and systems that started in the 1970s and continue to today, the U.S. is losing interest, and thus the intention to be a space power with the capability to lead in space is slowing. The National Security Space Strategy points out that, "Although the United States still maintains an overall edge in space capabilities, the U.S. competitive advantage has decreased as market-entry barriers have lowered. The U.S. technological lead is eroding in several areas as expertise among other nations increases."[5] This is the true threat to the American space program.

Due to the changing geopolitical situation after the Moon landing, the President and Congress decided that funding should be focused on other priorities. In the 1970s, the President and Congress cancelled the remaining three *Apollo* missions and instead focused NASA on building the first American space station, Skylab. There was a series of conversations between President Richard Nixon and his budget director, Caspar Weinberger, concerning this issue. Nixon wanted to cancel more missions and programs, specifically Skylab and the Space Shuttle; Weinberger, however, disagreed.[6] The cancelation of the final three Apollo missions were agreed upon and this led to the successful launch of Skylab in 1973, which, after only five short years in 1979, was brought down in order to focus on other projects. The ending of the Apollo program also ended the U.S. capacity for heavy lifting, by ending the construction of the Saturn V rocket. The next big project for NASA was the reusable space shuttle, which took several years to develop, during which time the U.S. had no access to space. It ran from 1981 to 2011. Tragedy struck in January 1986 when the Challenger exploded, but NASA was soon reinvigorated with a series of new space telescopes. Like other projects of the past, the shuttle project was deemed too costly and the program was ended. This was due to not only financial constraints but also the destruction of the Columbia shuttle. The Administration of George W. Bush tried to reinvigorate NASA with the Constellation program, which was designed to build a heavy-lifting-capability rocket similar to the Saturn 5, and establish a Moon base from which further missions to Mars and beyond would follow. The planned called for a new rocket to be in place and usable in 2015, return to the Moon by 2017, and start a permanent lunar base by 2020.

This program never received the funding it needed and was cancelled shortly after President Barack Obama took office in 2009. President Obama has tried to take the space program, again, in a different direction, by incentivizing the private sector to develop a reusable rocket. Currently, because the shuttle program has ended, the U.S. can only get into space aboard spacecraft flown by the Russian Federation, which charges the U.S. about $60 million per astronaut. Russia can therefore dictate the terms, opportunity, costs, and quite possibly influence U.S. space policy itself.

One of the pride and joys of the era of the space shuttle was the Hubble Space Telescope. This instrument has been unmatched in the knowledge that was gained by its discoveries. Unfortunately, it, too, has faced budget problems, and a debate erupted in the United States about its cancellation in 2009. It was designated for one last service mission and the life of the telescope was extended until 2015 when its replacement the James Webb Space telescope was to be launched. This again, however, may not happen, due to funding problems, as the James Webb launch date was pushed back to 2018, and there were serious discussions in Congress about cancelling it all together because it is too expensive.[7] In 2010, shortly after its completion, Congress seriously debated whether or not to continue funding the International Space Station and end it mission as planned in 2016, which would give the completed station just six years of completed operations despite having been constructed over the course of a decade, from 1998 to 2010.[8] After debate, however, funding was guaranteed until 2020, with a possible extension to 2026. The constant back-and-forth decision-making process makes long-term goals and plans extremely difficult, if not impossible. With no consistent long-term planning there is a rise of inefficiency and time wasted on starting and stopping.

This history is important because it shows the discontinuity and political football that has come to plague NASA and the U.S space program. The United States is currently without a heavy-lift capability that the Saturn V rocket provided and that the Aries rocket, under the Constellation program, would have provided. The United States is also without a reusable space shuttle, which the shuttle program provided and that the Orion capsule, under the Constellation program, would have provided. The Obama Administration's policy of shifting the responsibility of the space program to the private sector is being met with resistance by from retired astronauts. Former astronauts Neil Armstrong, Jim Lovell and Gene

Cernan stated in a letter to President Obama, "The availability of a commercial transport to orbit as envisioned in the President's proposal cannot be predicted with any certainty, but is likely to take substantially longer and be more expensive than we would hope."[9]

The severity of this threat is shown through the budget reductions and cuts to NASA since the mid-1960s. At its peak in 1966, NASA's budget was 4.4 percent of the federal budget. As of 2012 it is 0.48 percent of the federal budget.[10] This decrease in financial commitment has made NASA at times more efficient in terms of robotic missions, but has lowered U.S. progress in manned spaceflight and in doing so has lowered U.S. expectations and excitement about space. Armstrong, Lovell, and Cernan continue that, "Without the skill and experience that actual spacecraft operation provides, the USA is far too likely to be on a long downhill slide to mediocrity."[11]

Every technology needs to be upgraded and maintained if it is to work efficiently. Neglect will bring about failure on a number of levels. For example, without the U.S.'s own ability to determine for itself when a satellite needs to be repaired or upgraded because U.S. access is determined by another nation, it significantly decreases U.S. ability to repair technology. As the Commission to Asses United States National Security Space Management and Organization report states, "Space systems are vulnerable to a range of attacks that could disrupt or destroy the ground stations, launch systems or satellites in orbit. The political, economic, and military value of space systems makes them attractive targets for state and non-state actors hostile to the United States and its interest."[12] How is the U.S. able to counter these possibilities without our own access to space?

David R. Shedd, Deputy Director of National Intelligence for Policy, Plans and Requirements under the office of the Director of National Intelligence stated, "In addition to worrying about the next Pearl Harbor on Earth, we now must also consider the potential for a space Pearl Harbor."[13]

There is also the psychological vulnerability that is attached to this problem. The United States "won" the race to the Moon and defeated the Soviet Union—now the U.S. is dependent on Russia to get to and from space.

The United States is vulnerable to a major satellite failure. Satellites make for easy target for not only counterspace weapons, but cyber-attacks as well:

> The growing interdependence between U.S. civilian and military space systems further increases the likelihood that cyber-attacks might be launched against American commercial satellites, if for no other reason than military action directed against U.S. space capabilities will have to target the nation's broader space infrastructure to be successful. In addition, to po-

tential foreign adversaries seeking to avoid a direct military confrontation with the U.S. forces, whether a traditional uniformed military or "non-traditional" adversary (such as a terrorist organization), the commercial sector represents the "soft underbelly" of American space power, which can be attacked through cyberspace in such a way as to make determining the origin of the attack very difficult.[14]

Does the United States have the capacity to withstand a massive failure of, or attack on, its space systems? The short answer is yes; the United States as a constitutional republic can withstand a massive failure or space attack. However, that is not to say that there would not be significant challenges or trauma, such as domestic panic, rioting, and looting. Internationally, there may be political or even military moves on U.S. interests, but this would have to be a well-timed, organized attack. Financial systems could suffer from this lack of communication. The likelihood of such a scenario taking place is small. It is more likely to take place due to the growth in cyber-crime and cyber-espionage that an individual attack could damage or deem inoperable a military or communication satellite. There is a real concern about the U.S.'s current ability to withstand an attack in space.

There is another question of capacity that should concern the United States, however, and that is, not being attacked, but being outdone and surpassed by other nations. Should other nations such as China or Russia have enough time to catch up with U.S. abilities, they could very well surpass the United States and become the dominant space-faring nation. The United States itself would survive; however, in terms of prestige, being a global leader that people look to for inspiration could be lost for a generation. It is true that the United States eventually caught up with the Soviet Union and won the race to the Moon, but the times were different. The United States was locked in an intense ideological battle that had to be won on all fronts and at all costs. Today, the U.S. has been living peacefully with a communist China for over 40 years and with an autocratic/oligarchic Russia for the past 20 years. Should one of those nations be the first to develop a superior space technology or event, such as a Moon base or landing on Mars, the political culture of the United States today would be unlikely to rise to the challenge.

In order to stop the downward slide of the U.S. space program there need to be two crucial steps: (1) consistent unified leadership, not just from one President or one Congress, but from future leaders, both local and national; and (2) a solid commitment, again not from one President or one Congress, but from the nation as a whole. Why was President John F. Kennedy successful? He was an inspirational leader who received the commitment from a nation which, after his assassination, did not want to let him down. The same commitment was seen after the terror attacks of September 11, 2001. The United States was united and determined to eliminate al-Qaeda and its affiliates. While incomplete, significant damage has been done to al-Qaeda over the course of two Presidents and six Congresses. If the U.S. shows the same determination in the areas of space, it will be extremely successful.

The United States today faces many challenges domestically and globally. However, with that grows a threat and risk to the U.S. space program. The program is suffering from a failure of domestic leadership from multiple administrations and multiple congresses that demand excellence but refuse the resources to achieve it. There has been the consistent starting, stopping and changing of space policy that has left the industry frustrated and confused. A tremendous amount of time and resources, both financially and with the labor that was put into the projects have been wasted through this process. Due to this, there is a domestic threat to the U.S. space industry from this lack of leadership. This is directly putting at risk the space program to both being vulnerable to an attack and being surpassed by a rising new power. Should the U.S. fail to rise to the challenges and correct the course it has taken, a devastating event, such as a "space Pearl Harbor" could have damaging effects to U.S. national security and interest. In addition, the rise of another nation that surpasses will be a serious blow to American prestige and inspirational influence in the world. No nation is inspired by the second place finisher, nor does it follow its example in other areas. Over time, with continued setbacks, this could turn into a culture that will no longer be inspired by what America can accomplish but what other nations under different political systems can accomplish. This can be seen historically during the race to the Moon in the Cold War. When Yuri Gagarin made the first successful flight into space, Egypt's President Nasser was in awe of the "gigantic scientific capabilities of the Soviet people and had no doubt that the launching of man into space will turn upside down not only many scientific views, but also many political and military trends."[15]

Attempts at leadership have appeared from time to time. President Reagan in his speech to the nation after the *Challenger* disaster stated, "We've grown used to wonders in this century. It's hard to dazzle us. But for 25 years the United States space program has been doing just that. We've grown used to the idea of space, and perhaps we forget that we've only just begun. We're still pioneers. They, the member of the *Challenger* crew, were pioneers."[16] Americans are still just pioneers, very similar to the navigators during the age of exploration in the 16$^{th}$ century. As President Reagan correctly alluded to in that same speech, "On this day 390 years ago, the great explorer Sir Francis Drake died aboard ship off the coast of

Panama. In his lifetime the great frontiers were the oceans, and a historian later said, 'He lived by the sea, died on it, and was buried in it.' Well, today we can say of the *Challenger* crew: Their dedication was, like Drake's, complete."[17]

The U.S. as a nation has grown used to being able to work in space; however there is still so much to do. The U.S. needs to start allocating the resources not so that Americans may be simple pioneers, but that they can learn enough and begin to master the environment of space, as they do on land, at sea, or in the air. It is unfortunate that, if the U.S. continues the course it has taken over the past several decades, the tortoise will catch up with the hare.

## Notes

1. Palazzo, Steven M. (R-MS), Chairman Subcommittee on Space and Aeronautics, April 26, 2012, http://science.house.gov/sites/republicans.science.house.gov/files/documents/HHRG-112-SY16-WState-P000601-20120426.pdf

2. Office of the Director of National Intelligence National Security Space Strategy 2011 p..i.

3. Palazzo, April 26, 2012

4. Kennedy, John F. Rice University September 12, 1962, http://er.jsc.nasa.gov/seh/ricetalk.htm.

5. National Security Space Strategy 2011 p. 3.

6. Weinberger, Caspar. Memorandum to the President of the United States. Office of Management and Budget. August 12 1971 http://www.alternatewars.com/SpaceRace/Speeches/Nixon_NASA_Budget_71.htm

7. Peter Pachal, "What We Could Lose if the James Webb is Cancelled," July 8, 2011, http://www.pcmag.com/article/print/266679.

8. Joel Achenbach, "As Space Station Nears Completion, It Faces End of Mission," *The Washington Post*, July 13, 2009, http://www.washingtonpost.com/wp-dyn/content/article/2009/07/12/AR2009071201977.html (accessed April 8, 2013).

9. Armstrong, Lovell and Cernan.

10. United States Government. Federal Budget of the United States. United States Government Printing Office. 2012 http://www.gpo.gov/fdsys/search/pagedetails.action?packageId=BUDGET-2012-BUD

11. Armstrong, Lovell, and Cernan.

12. Commission to Assess United States National Security Space management and Organization 2001, p. 12.

13. Shedd, David R Office of the Director of National Intelligence. June 9, 2010, http://www.dni.gov/speeches/20100609_speech.pdf

14. Christopher M. Petras, "The Use of Force in Response to Cyber-Attack on Commercial

Space Systems," *Journal of Air Law and Commerce*, Vol. 67, No. 4 (Fall 2002): 1213–1268.

15. Walter A. McDougall, *The Heavens and the Earth: A Political History of the Space Age* (New York: Basic Books Press, 1985), p. 287.

16. Ronald W. Reagan, address to the nation, January 28, 1986.

17. Reagan Ibid .1986

Made in the USA
Monee, IL
04 November 2026

67380504R00059